What Science Is and How It Really Works

Scientific advances have transformed the world. However, science can sometimes get things wrong, and at times, disastrously so. Understanding the basis for scientific claims and judging how much confidence we should place in them is essential for individual choice, societal debates, and development of public policy and laws. We must ask: What is the basis of scientific claims? How much confidence should we put in them? What is defined as science and what is not? This book synthesizes a working definition of science and its properties, as explained through the eyes of a practicing scientist, by integrating advances from philosophy, psychology, history, sociology, and anthropology into a holistic view. Crucial in our political climate, the book fights the myths of science often portrayed to the public. Written for a general audience, it also enables students to better grasp methodologies and helps professional scientists to articulate what they do and why.

JAMES C. ZIMRING is a professor of pathology at the University of Virginia. The recipient of many awards for his research and teaching, he is recognized as an international expert in the field of transfusion biology and routinely delivers academic lectures both nationally and internationally.

What Science Is and How It Really Works

JAMES C. ZIMRING
University of Virginia

CAMBRIDGE
UNIVERSITY PRESS

Shaftesbury Road, Cambridge CB2 8EA, United Kingdom

One Liberty Plaza, 20th Floor, New York, NY 10006, USA

477 Williamstown Road, Port Melbourne, VIC 3207, Australia

314–321, 3rd Floor, Plot 3, Splendor Forum, Jasola District Centre, New Delhi – 110025, India

103 Penang Road, #05–06/07, Visioncrest Commercial, Singapore 238467

Cambridge University Press is part of Cambridge University Press & Assessment, a department of the University of Cambridge.

We share the University's mission to contribute to society through the pursuit of education, learning and research at the highest international levels of excellence.

www.cambridge.org
Information on this title: www.cambridge.org/9781108476850
DOI: 10.1017/9781108569149

First published 2019 (version 3, August 2022)

Printed in Great Britain by Ashford Colour Press Ltd., August 2022

A catalogue record for this publication is available from the British Library

Library of Congress Cataloging-in-Publication data
Names: Zimring, James C., 1970– author.
Title: What science is and how it really works / James C. Zimring (University of Virginia, Charlottesville).
Description: Cambridge ; New York, NY : Cambridge University Press, 2019. | Includes bibliographical references and index.
Identifiers: LCCN 2018058439 | ISBN 9781108476850 (hardback : alk. paper) | ISBN 9781108701648 (pbk. : alk. paper)
Subjects: LCSH: Science–Methodology. | Reasoning.
Classification: LCC Q175.32.R45 Z56 2019 | DDC 501–dc23
LC record available at https://lccn.loc.gov/2018058439

ISBN 978-1-108-47685-0 Hardback
ISBN 978-1-108-70164-8 Paperback

To my wife Kim and daughter Alex

Contents

Acknowledgments

I am delighted to acknowledge the long list of people to whom I am indebted for all of their help, support, encouragement, and critical feedback. The list is sufficiently long that I will inevitably omit someone whom I need to thank; if you are that person, please accept my appreciation and apology. It should not be taken that those whom I acknowledge necessarily hold my views; my goal is to express my thanks without making them involuntarily complicit.

I have heartfelt gratitude for those who have been my scientific teachers and mentors, who patiently supported and guided my intellectual development, and without whom I would never have experienced science at all. These include Michael Carmichael, Dennis Liotta, Margaret Offermann, Judith Kapp, Christopher Hillyer, Tristram Parslow, James AuBuchon, and others. Your kindness, patience, and guidance have been a *sine qua non* of my development. A special thanks to David Jollow for patiently guiding my development for 30 years.

I would be remiss not to acknowledge the debt I owe to the students who took my course on this topic, first at Emory University and later at the University of Washington. As is typically the case in teaching, I likely learned as much, if not more, from my students than they learned from me. I am, and shall remain, grateful for all the critical debate and intellectual engagement from a truly wonderful group of young and energetic minds. I hope each of you have found or are finding the scientific careers and pursuits that you seek.

I have a special and heartfelt thanks to give to each of the students, fellows, staff scientists, and lab members with whom I have worked in the labs that I have been a part of or that I have overseen. I have shared scientific journeys with each of you, navigating ever-changing webs of belief, which has been the joy of my professional life. I give special thanks to Krystalyn Hudson, Seema Patel, Maxime Desmarets, Justine Liepkalns, Chantel Cadwell, Matthew Schniederjan, Arielle Medford, Linda Kapp, Rene Shaw, Brian Harcourt, Jennifer Harcourt, Gloria Kelly, Florence Roan, Kyle McKenna, Travis Spurling, Christopher Gilson, Kate Henry, Geetha Mylvaganam, Alice Long, Nikki Smith, Ashley Bennett, Traci Chadwick, Suzanna Schott, Kathryn Girard-Pierce, Jeanne Hendrickson, Sean Stowell, Hayley Waterman, Karen de Wolski, Heather Howie, Ariel Hay, Jacqueline Poston, Xiaohong Wang, and Kimberly Anderson.

To my professional friends and colleagues, who have patiently tolerated my incessant babble on the topic of this book and who have continued to support and encourage me, I do not have the words to express my profound gratitude. There are many of you and I appreciate you all; in particular, I would like to thank Steven Spitalnik, Patrice Spitalnik, Eldad Hod, David Masopust, Vaiva Vezys, Jaap Jan Zwaginga, Gary Falcon, Theresa Prestwood, Zoe Falcon, Suzanna Schott, Taylor Maxson, Angelo D'Alessandro, Carla Fowler, Dana Devine, Krystalyn Hudson, Melissa Hudson, Richard Francis, Tiffany Thomas, Jeanne Hendrickson Jill Johnsen, Stephanie Eisenbarth, and John Luckey.

I would like to express my extreme gratitude to those who contributed to the editing, development, and production of this manuscript. My thanks to Susan Beckett, Robert Swanson, and especially Al Desetta for his expert help in editing several versions of this book. I give my special thanks to members of Cambridge University Press (Katrina Halliday, Anastasia Toynbee, Aleksandra Serocka, and others who have so graciously accepted and developed this project), and to the anonymous peer-reviewers whose identities I shall never know but whose comments and criticisms were essential and very much appreciated. I especially thank Dr. Steven Spitalnik, Dr. Eldad Hod, Dr. Jacqueline Poston, Dr. Jeffrey Kasser, and Dr. Paul Humphreys for their insights and comments on a number of components of the text.

Most importantly, and above all else, I would like to give my everlasting love and gratitude to my family. To my mother and father for literally making me who I am and always supporting and encouraging my life's journey, to my wife Kim for taking that journey with me with endless love and support, and to my daughter Alex, the light of my life, who reminds me every day of the wonder that our world holds. To my brother Jon for a lifetime of debate and laughter, to Michelle, Matthew, Joshua, Kayla, David, Margaret, John, and Nicole for everything you are.

James C. Zimring

Note added on proof:
While this book was already in press, and the final typesetting had been done, a new book was published by Dr. Lee McIntyre entitled "The Scientific Attitude: Defending Science from Denial, Fraud, and Pseudoscience." In this work, Dr. McIntyre advances novel concepts and defines the attitude by which scientists approach evidence as a necessary component of science, which can help to defend it against deniers and pseudoscientists. Had "The Scientific Attitude" been available to me while this book was still being written, I would have certainly referenced it highly, especially in sections discussing how scientists handle data and the general property of attempting to mitigate error as a characteristic of science. These ideas are fully developed in Dr. McIntyre's book that was published six months prior to the current work.

Introduction

THE IMPORTANCE OF DEFINING SCIENCE

The terms "science" and "scientific" have come to have a special meaning and to carry a special weight in modern society. Professional scientists tell us that genetically modified foods are safe to eat, that industrial emissions are causing global warming, that vaccines don't cause autism, and that some medications are safe and effective while others are not. A consumer product seems more trustworthy if it's described as "scientifically proven" or if "clinical studies have demonstrated its effectiveness." Politicians and lobbyists often evoke "scientific proof" in arguing for certain positions or policies. Our federal government invests taxpayer dollars in "scientific research" of different varieties. Whether something can be categorized as "science" determines if we allow it to be taught in our public school science curricula, as in the ongoing debate over teaching evolution vs. intelligent design theory.

To evoke "scientific" in the description of a claim makes it seem different, and likely more credible, than claims of a nonscientific nature. Rarely, if ever, do we read of a healthcare product being "philosophically proven to heal" or "theologically demonstrated to cure." But why does the label "scientific" carry any special weight at all? What does it mean when we say that something is "a scientific fact"? Doesn't science sometimes get things wrong, and if so, why should we believe future scientific claims when past claims have sometimes been in error? Even if we stipulate that science does and should carry special weight, how are we to know if something really is scientific? How much trust should we put in the claims that scientists make, and how are we to evaluate whether the claims themselves really are based on science? Overall, the question becomes: What does it mean to be scientific, and how can we define science?

Scientific and technological progress has transformed how humans live in countless ways. For this reason, science's ability to predict and manipulate the natural world certainly appears to imbue scientific knowledge with at least a greater degree of practical utility than other forms of knowledge, if not a greater degree of truth. Most people seem to accept that science is different in some way, and if such is the case, it seems as though we should be able to define what science is and how it is different from that which is not science.

As for myself, I come to this problem as an academic professional physician and scientist who pursues both independent research and is deeply involved in the teaching of students of science at the graduate level. The formal education of scientists focuses almost exclusively on the doing of science, not the understanding of what science is. Most doctoral students who are in advanced educational programs designed to train the next generation of scientists spend their time observing mentors and fellow students and, initially, learning through imitation and repetition. Formal coursework is provided in order to instruct students in the current beliefs and cutting-edge theories of their chosen field and the accepted evidence that supports (or refutes) such theories; however, little if any classroom instruction is typically given regarding the *process* by which science is done. Rather, this is learned by doing and by observing the doings of others (fellow students, technicians, postdoctoral fellows, and faculty mentors), over and over again. Eventually, one may begin to see patterns in how the field functions. One often begins to think about the process of the scientific approach itself. One analyzes *how* humans study nature, the strengths and weaknesses of different approaches, and the pitfalls to be avoided. However, in this latter task, it is ironic that while scientists themselves are the leaders in the practice of science – the ones who know "how to do the thing" – they are amateurs in the formal analysis of what science is.

There are whole academic fields (separate from the practice of science itself) that focus precisely on how scientists study nature; in particular, the philosophy, history, psychology, anthropology, and sociology of science, as well as integrated studies that combine these different areas to give a broader view. Whereas my fellow scientists and I study nature, these fields study us. Sadly, in my experience, practicing scientists are seldom very much aware of the details of such fields and in many cases may not know that such fields even exist, other than having heard rumors to that effect. This is not to say that scientists aren't very good at practicing science; theoretical knowledge of the inner workings of an internal combustion engine and the ability to drive a car are separate areas of understanding, and one can be an expert in one while being entirely ignorant of the other. Many scientists can recognize good science when they see it and can call out flawed science when it is encountered, or at least they believe they can, as this is the basis for the entire peer-review system in science. However, because they can judge science does not necessarily mean they can clearly articulate the theoretical underpinnings of what defines science, of how science works and/or how science fails.

If even professional scientists are not typically trained in the general underpinnings of scientific knowledge claims, how is there any hope for nonscientists to understand what level of confidence they should (or should not) place in science and the claims it makes. Moreover, how can people be expected to distinguish valid scientific claims from all manner of flim-flam and fluff? The goal of this work is to help lay people, students of science, and professional scientists understand and explain how science works in general, the strengths and weaknesses of scientific thinking, and the extents and limits of scientific knowledge claims.

Despite the weight that the label of science may carry with many people, it is an utter fiction that there is (or ever has been) a uniform consensus among scientists (or anyone else for that matter) as to what precisely defines science. This question has been tackled over the years by many great scholars and yet there is not a clear and unequivocal answer. Nevertheless, much progress has been made, and this has generated a greater understanding of characteristics of science, its practice, and its strengths and limitations. The goal of this work is to communicate a broad view of that progress. This is an ambitious goal to be sure, but the difficulty of the task does not diminish its importance. What has been learned is surprising, counterintuitive, and complex. Ultimately, it speaks not only to science but to the human condition itself.

GOALS AND ORGANIZATION OF THIS WORK

When I introduce the reader to concepts generated by the outside fields that study science itself, I am reflecting the insights, innovations, and contributions of others – standing on the shoulders of giants. I name the giants when I can and have taken care to try and point out to the reader what the original source of many concepts are and what resources and further reading one might do to explore the more granular details of different specialized areas of focus. However, the richness of these different fields goes so deep, that much will be neglected – other works devoted to the finer nuances and details of each component field are abundant, and one need only seek them out. Herein I attempt to synthesize key ideas into a unified framework, hopefully making it coherent to the reader. In addition to explaining the progress of those who study science, I contribute the perspective of the thing being studied – a view of great utility. I speak with the voice of the bacterium on the observations, interpretations, and theories of the microbiologist, for I believe that I know what it is to grow in the chaotic ferment of the microbial culture.

The book is organized into three parts. In Part I (Chapters 1–3), the individual working parts of scientific reasoning and logic are described (and then an attempt is made to draw a picture of scientific reasoning as a whole). In Part II (Chapters 4–8), flaws that undermine natural human observation, perception, and reasoning will be described. In Part III (Chapters 9–13), I will explore how scientific processes and methods try to address these flaws, attempting a distinction between scientific and nonscientific thinking. An over-arching theme of the final part of the book is how science mitigates the tendencies of normal human thinking to "get the world wrong" in particular situations.

The first goal of this book is to help guide nonscientists in having reasonable expectations of what science can and can't do. Scientific claims are often regarded with either too much confidence or too much skepticism by different groups of the general public. This book strives to lay the groundwork for a healthy balance in how to weigh scientific knowledge claims. The second goal of this book is to help professional scientists gain a better understanding and codification of the strengths and weaknesses of their craft and the role they play in portraying it. Of high importance to this latter audience is the recognition that it is quite intoxicating, from an ego standpoint, for scientists to be regarded as the arbiters of "true" knowledge. This has been described as "the Legend" of science.[1]

The extreme version of the Legend claims that "science aims at discovering the truth, the whole truth, and nothing but the truth about the world" – a less grandiose version of the Legend states that science is "directed at discovering truth about those aspects of nature that impinge most directly upon us, those that we can observe (and, perhaps, hope to control)." While it is argued that the Legend has been abandoned by those who study science, the Legend (or a slightly weakened version of it) seems very much alive among some in the lay public. In my experience, scientists themselves hold onto a version of the Legend, and while it is less extreme than a philosophical truth, it nevertheless has some component of being "truer" than that which is not science. In my view, scientists should neither seek nor accept the extreme versions of the Legend, which are pleasant in the short term but harmful in the long term, and ultimately destructive as they lead to unsupportable claims. Failures to live up to hyperbolic

---

[1] Kitcher P. 1995. *The Advancement of Science: Science Without Legend, Objectivity Without Illusions*. New York: Oxford University Press.

attributes only leads to anti-Legend, those who claim with vitriolic hostility that science is, at best, nothing at all, and at worst, a grand conspiracy to dupe the world. Rather, we must seek a balanced and honest view based on realistic assessments. Science's greatest apparent weaknesses are, in actuality, its greatest strengths; as professional scientists we should embrace this and not seek to minimize or ignore it. In the greatest traditions of scientific scholarship, let the existing data of what science is inform us as to the properties of science itself – let us look it in the eye, unflinching, and without spin or propagandist inclinations.

A RIDICULOUSLY BRIEF HISTORY OF SCIENCE VS. NONSCIENCE

Early on (dating back to antiquity) and arguably from a position of great overconfidence, scholars of science often stated that science (or natural philosophy as it was called before the 1830s) dealt with facts, whereas other schools of thought dealt with opinions. However, as many established scientific facts were later rejected by subsequent generations, they came to be understood as fallible and thus not so different from opinions.[2] Yet the realization that scientific facts are imperfect doesn't mean that they don't have a different character than nonscientific knowledge claims. But if they do (which is not a given), why is that so, and what is the justification for such a view? Later thinkers gravitated to the notion that if it is not fact that distinguishes science from other ways of knowing, then it must be the manner by which scientific claims are generated and/or evaluated that distinguishes science from nonscience. In other words, the method that science uses to create knowledge has a special character that is different from nonscientific approaches.

---

[2] Laudan L. 1983. "The Demise of the Demarcation Problem." In Cohen RS, Laudan L (Eds.). *Physics, Philosophy and Psychoanalysis: Essays in Honor of Adolf Grunbaum.* pp. 111–27. Dordrecht: D. Reidel.

For the reasons just stated, most modern attempts at defining science have focused on *methods* or *modes of thinking* that distinguish scientific activities from nonscientific activities rather than the specific content of scientific knowledge claims. However, while one often encounters discussions of "the scientific method" and its application to investigation, there is a lack of agreement about what precisely this method entails, and there are those who argue that the very notion of a scientific method is itself an utter myth.[3] It has further been argued that different areas of scientific study favor different types of method(s), and thus one cannot precisely define "science" or "the scientific method" per se.

Moreover, it has been argued that even if some broader characteristics can help identify a method as scientific, precisely demarcating how science differs from other ways of thinking is neither possible nor useful.[4] It has even been claimed that science flourishes only with a distinct lack of required methodology, and that attempts to codify a scientific process will only serve to destroy it – in other words, the only rule of science is that "anything goes."[5] However, this latter view is somewhat radical and is certainly not embraced by most professional scientists, as evidenced by certain generally agreed-upon standards used in the practice of peer review of reports of scientific discoveries, grant applications, and research.

Scholars of science have often rejected any definition that would render the great historical scientists as "nonscientific." While this seems logical, it presupposes that those who have made the most progress and achieved the most recognition were those acting most scientifically – a question we shall explore in detail. Perhaps more importantly, it assumes that the scientific method, however we define it, has been stable over time, a claim that seems hard to justify.

[3] Bauer HH. 1992. *Scientific Literacy and the Myth of the Scientific Method.* Urbana and Chicago: University of Illinois Press.

[4] Laudan L. 1983.

[5] Feyerabend P. 1975. *Against Method: Outline of an Anarchist Theory of Knowledge.* New York: New Left Books.

What "scientific" means in 2019 may be very different from what it meant in 1919, 1819, or 1719. This doesn't mean that there aren't common threads that can be woven into a definition, and we shall endeavor to identify those threads. But the idea that universal factors must be present in all science over time – that science itself is not evolving – is a difficult position to support. But if science is evolving over time, is it a clearly definable thing? This, too, will be addressed.

Even if a clear and universally accepted distinction between science and nonscience that can categorize each and every instance does not exist, this does not mean that there is no difference between science and nonscience; the presence of gray does not eliminate the distinction between black and white. Insisting that no definition of science can be put forth unless it is perfect, identifying necessary and sufficient conditions, with no ambiguity or unclear instances, falls into the trap of black-and-white thinking (a.k.a., the perfect solution fallacy). In most cases, the world does not come in black and white, and attempts to force it into yes/no categories fails because the world is a continuum encompassing all shades. Nevertheless, even imperfect definitions can be both real and useful. In the translated words of Voltaire, "the perfect is the enemy of the good." For these reasons, ongoing examination of the black, the white, and the gray areas of this topic remains necessary.

More recently, a number of scholars have analyzed the definition of science with the recognition that absolute categories can simply be an artifact of language and human thinking and that previous failures to define science were inevitable unless one treats categories as more fluid and with boundaries that are less sharply defined.[6] It has been suggested that science vs. nonscience must be analyzed using looser boundaries, with families of properties or "cluster analysis."[7] Difficulty in categorization is by no means unique

---

[6] Dupré J. 1993. *The Disorder of Things: Metaphysical Foundations of the Disunity of Science*. Cambridge, MA: Harvard University Press.

[7] Mahner M. 2013. "Science and Pseudoscience: How to Demarcate after the (Alleged) Demise of the Demarcation Problem." In Pigliucci M, Boudry M (Eds.). *Philosophy of*

to science, it is a regrettable problem of language and thought and afflicts many areas – philosophers and linguists can at times have a very hard time precisely defining things that are nevertheless agreed to exist. Still, this in no way undermines the importance of refining definitions and continuing to characterize and describe. Defining what science is (and is not) remains a task of great importance.[8]

THE ILLUSION OF SCIENCE: A VERSION OF THE LEGEND LIVES ON

Science is often presented and perceived as a logical and orderly process that makes steady progress in understanding nature. Scientific presentations and publications are viewed in this fashion. Textbooks describe how seminal experiments were carried out to challenge scientific ideas from the past and how theories were adjusted to encompass new and surprising results. Scientific findings are reported with the appearance of being rational and logical as fields march steadily forward to better theories and greater understanding. Moreover, scientific beliefs are often stated as unequivocal facts. In 2012, when observations were made, consistent with what would be predicted if the Higgs boson existed, most reports didn't claim to have "encountered evidence consistent with the presence" of a Higgs boson; rather, it was stated that the Higgs boson had been "discovered"! In actuality, a scientific fact is nothing more than that which has stood up to rigorous testing thus far by the scientific methodologies currently available, but this is not how scientific facts are often presented.

Pseudoscience: Reconsidering the Demarcation Problem. Chicago and London: University of Chicago Press, pp. 29–43.

[8] For a review of the demarcation issue and its history, see Nickles T. 2006. "The Problem of Demarcation." In Sarkar S, Pfeifer J (Eds.). *The Philosophy of Science: An Encyclopedia. Vol 1*. New York: Routledge, pp. 188–197. See also Nickles T. 2013. "The Problem of Demarcation, History and Future." In Pigliucci M, Boudry M (Eds.). *Philosophy of Pseudoscience: Reconsidering the Demarcation Problem*. Chicago and London: University of Chicago Press, pp. 101–120.

In order to understand science, it is necessary to jettison the unrealistic hyperbole that has been mistakenly assigned to it (from a number of sources), and that has been perpetuated by practicing scientists and science enthusiasts. The flaws of science need to be called out, and greater attention must be directed to its problems, weaknesses, and the limits of what it can show us. We must turn the scientific microscope back on itself and dissect the specimen with a critical and analytic eye, without succumbing to the tendency to give descriptions that are unjustifiably favorable. By describing science as it really is, warts and all, we can simultaneously view science realistically and more accurately differentiate it from other ways of thinking. That science is imperfect and flawed doesn't mean that it isn't distinct from other knowledge systems or that it can't be described, if not defined. Likewise, its flaws and imperfections don't prevent it from being the most effective means (thus far) of exploring and understanding nature. Just as democracy may be "the worst form of government except for all the others,"[9] the same could be said for science's role in understanding the natural world. The goal of this book is to view science more realistically, not to make vain and misguided attempts to defend a grandiose view that is out of step with the actual entity.

The depiction of science as a logical and orderly process, governed by a specific method and leading to firm facts about nature, although regrettably a distortion of how science is really carried out, is the byproduct of how scientific findings are communicated among scientists. The reasons for this will be expanded upon later, but for now it's important to note that while such distortions may be necessary to communicate scientific findings efficiently, it is profoundly damaging to present this illusion of science rather than the reality of how it is practiced. Practicing scientists typically understand that the distortion does not reflect the reality of the situation. However, those

---

[9] This quote is attributed to Winston Churchill, but apparently he was quoting an earlier source that remains unidentified.

outside a field (or outside science altogether) may miss this distinction, believing science to be other than it is. Indeed, this distortion has likely contributed to the genesis of the Legend in the first place. In trying to understand science we can mistake the mirage for the desert. We must be willing to accept that the tempting oasis is merely an image, and we must focus instead on understanding the desert itself, which is the reality of the situation.

MISSING THE FOREST FOR THE TREES

Attempts to describe and understand science have often focused on its component parts, which is a necessary process of any deep analysis of an entity. Efforts have been made to distinguish science from nonscience based on logical constructs, the sociology of science, the psychology of science, and the history of science. However, while each of these areas plays a central role in what it means to practice modern science, none of them tell us the whole story. Trying to understand science exclusively through analysis of its parts is like the ancient story of the three blind humans each studying a different part of an elephant.[10] The person feeling the legs may think it's a tree, the person feeling the tail may think it's a rope, and the person feeling the trunk may assume it's a snake. Each is correct in their observations, but to understand what an elephant really is requires a broader view that merges the component parts into a greater system. This has been recognized in recent decades, and academic disciplines that attempt to generate an overall synthesis of what science is and how it works have emerged.

Modern science is a combination of multiple working parts, including: advanced instrumentation and approaches to observation, human perception and cognition, computational analysis, the application of logic and reasoning, and the effect of social bodies on how

---

[10] This story can be found at least as far back as the Buddhist text *Udana 6.5* from the middle of the first millennium BCE (and likely earlier) (https://en.wikipedia.org/wiki/Blind_men_and_an_elephant).

investigation is conducted and interpreted. To grasp modern science, each of these factors must be accounted for; we need to understand both the forest and the trees, as neither has its fullest meaning without the other. In the last century alone, technologies and methodologies have greatly increased the scope of what we can observe (and also misinterpret) in ways never before possible. Cognitive psychology has taught us much about common errors in human reasoning, perception, and observation. Computational capacity allows us to generate and analyze previously overwhelming volumes of data and to make comparisons and analyses far beyond the capacity of the human mind and, in doing so, to lead to new errors an individual human mind would have a hard time making. Statistics has made great strides in its ability to analyze levels of error and to quantify uncertainty, to determine the nature of underlying mechanisms by the distribution of data, and to evaluate associations. Philosophers of science and logic have provided us with a much clearer understanding of the strengths and shortcomings of reasoning than ever before, as well as novel insights into the nature of evidence and the extent to which one can actually verify or reject an idea. Sociologists and anthropologists have learned a great deal about the effect of group dynamics and scientific societies on thinking. Linguists and philosophers of language have identified sources of ambiguity and miscommunication. To understand the current limits of science, each of these must be examined.

The understanding that science is a complex machine, with multiple working parts that need to be understood both individually and in aggregate, is essential. One cannot understand how an internal combustion engine works as a whole without understanding the function of a spark plug. Yet the full implications of what a spark plug does are unintelligible without a preexisting understanding of the entire engine. To break into a circle of codependent knowledge such as this, one may have to visit, and then revisit, the parts and the whole. In this book, the different individual components of science are described first. Later chapters then illustrate the interactions of individual parts as a system. For this reason, the reader is encouraged to loop back to earlier parts of the book, if needed, during the development of the narrative. In this way, the full implications of the properties of the individual parts of science may become clearer in later sections, when the whole system is described.

SCIENCE IS AN EXTENSION OF HUMAN THINKING THAT ALSO VIOLATES ASPECTS OF HUMAN THINKING

Science is often portrayed as something very different from normal human behavior or thinking. It seems fair to suspect that science is indeed distinct from normal human thought in at least some ways; after all, most humans are not scientists. However, just because science may differ from typical thinking in some fundamental ways doesn't mean that it is entirely foreign to human cognition. Rather, only a very small number of differences between science and normal human thinking distinguish the two. This may explain part of the difficulty in attempting to demarcate science from nonscience. Because they are so closely related, it's easy to point to apparent exceptions that violate any potential distinction. In other words, defining characteristics attributed to science have been rejected by some precisely because it is easy to find the same characteristics in thinking that is agreed to be nonscientific. Likewise, thinking that is ubiquitous in nonscientific pursuits has been easy to identify as an important component of methods of practicing scientists. Large categorical differences between science and nonscience are not to be found; rather, the small differences between them hold tremendous weight but can also be difficult to pin down.

The differences between scientific and normal thinking, while small, are nevertheless both fundamental to science and also deeply baked into normal human cognition. This in part explains why scientists must undergo so much formal training; they must first become aware of certain normal human tendencies and then learn how to manage and/or overcome them. One must learn how to ignore certain parts of what it is to think like a human – a difficult task for a thinking human to accomplish. While humans have been on this Earth for a long time, modern science has only been an activity of ours for about 400 years. The generation of science has been a development of human understanding, not of biology – prehistoric humans had essentially the same brains we do today but didn't have advanced science and technology. Scientists are engaging in a learned practice no different than any technique or skill that is developed and refined over time. But the reasoning and thinking that has been learned have subtle but essential parts that are different than our natural (or traditional) reasoning, otherwise we would have had science all along. Exploration of these small differences is key, as well as developing an understanding that because they may violate our natural thinking they feel "wrong" or "counterintuitive," while actually being quite correct.

DEFLATING SCIENCE TO A REALISTIC AND THEREFORE DEFENSIBLE ENTITY

It is an aim of this work to deflate common scientific hyperbole to a realistic and therefore more defensible description. Despite its accomplishments, the ability of science to predict nature is always limited in multiple ways. Scientific predictions and conclusions can never be certain, never be perfect, are certainly never infallible – nothing is ever "proven"' in a formal sense of the word. Even things that science labels "Laws of Nature" are themselves reversible if later understanding arises that requires their modification or even wholesale rejection. Ironically, when combined with its other properties (and this is key), it is precisely the recognition of its own limits and imperfections and the practices to which such recognitions then give rise, that constitutes science's greatest strength.

There are many systems of belief that provide much better explanations of experience than does science. Indeed, some systems can explain why anything and everything occurs; science makes no such claim, and those who would suggest that science currently has these ambitions are misguided. Other systems often claim to know absolute truths; modern science does not, and in this way the Legend truly is dead. If a conceptual framework by which you can comfortably explain the whole world and all experience is your goal, if you are uncomfortable (or even just don't favor) wrestling long term with ignorance and confusion,[11] then science is not the instrument for this. Why then, you might ask, should one embrace science over other systems that claim to provide truth and explanation? The answer is that if the ability to predict and control nature is your goal, then the very science that fails to explain everything with the comprehensive certainty of other systems of belief outperforms everything else every day of the week and twice on Sunday. It is understanding how this flawed system can repeatedly make uncanny and correct predictions better than any other approach to knowledge yet described and to the opposite of what normal human inclinations and explanations would do – that is the goal of this work.

---

[11] In the book, *Ignorance,* a compelling argument is made that science focuses on ignorance more than knowledge – a key component that manifests itself in much of scientific practice. Firestein S. 2012. *Ignorance: How It Drives Science*. Oxford, UK: Oxford University Press.

PART I

1 The Knowledge Problem, or What Can We Really "Know"?

And since no one bothered to explain otherwise, he regarded the process
of seeking knowledge as the greatest waste of time of all.

– Norton Juster, *The Phantom Tollbooth*

GENERAL DESCRIPTION OF THE KNOWLEDGE PROBLEM

Francis Bacon, one of the luminaries of modern science, is thought to
have said that "knowledge is power." Since Bacon made that statement,
it has become abundantly clear that humans have a very distinct and
difficult "knowledge problem." There is a fundamental defect in how
we come to know anything, and while this is recognized as a problem,
the depths of the problem are seldom appreciated and even less
frequently discussed. At first glance such a statement may seem ridiculous. What is the problem in saying someone knows something? I know
where I am and what I'm doing. I know the names and faces of my
friends, family, and acquaintances. I know how to drive a car, how to
cook (at least somewhat), and how to pay bills. In fact, just to navigate
the tasks of daily life one has to "know" a great number of things.

The knowledge problem in its classic form is not a challenge to
one's knowledge of the things that one has observed or the techniques
one has acquired. No one questions that you know that you have a
car, that you know you're married, or that you know you own a
collection of Elvis Presley commemorative plates from the Franklin
Mint that you inherited from your grandmother.[1] However, the word

---

[1] Much time and energy has been spent in classic philosophy debating whether we can
actually know anything of the external world. However, in everyday life, it is
generally accepted that our experience is the result of some external reality that is
actually out there.

"knowledge" takes on a very different character as soon as one goes beyond that which is directly observed or experienced. Substantial problems emerge the moment that knowledge claims are extended to things that have not yet been observed, either now or in the past or future. An additional and separate problem emerges once one claims knowledge regarding the relationships and associations between things. Over the past two millennia, as historians, sociologists, anthropologists, and philosophers have analyzed how claims to knowledge arise, develop, and collapse, there has been an expanding appreciation of just how limited our ability is to know.

Since ancient times humans have been on a quest for a higher form of knowledge that can make universal claims. Knowledge in its most ambitious form consists of fundamental truths about which we can be certain, about which we cannot be mistaken or wrong, of which we are sure. Facts and understanding of this kind can be forever considered true; we no longer need to worry about their validity, as these are things that must be so. We can put them in the "true folder" on our computer and forget about having to continue questioning them, they are certain. This is the meaning of knowledge in its extreme form, and it is with this form of knowledge that the knowledge problem is most pronounced.

For some people, it is both unacceptable and distasteful to admit that there is no certain knowledge; they hold particular ideas and convictions with absolute certainty, and there are many systems of belief constructed on such premises. For others, certain knowledge does not and need not exist, as it is not required to navigate the world and enjoy one's life.[2] From a pragmatic point of view, if an understanding works, then it is useful, even if in substance it only reflects some misunderstanding. In reading this book, I would ask those who believe in certain knowledge to have an open mind when we analyze

---

[2] Even for skeptics of knowledge, many would hold that forms of mathematics and logic constitute certain knowledge; the potential problems and limitations of this view are discussed later.

the basis of its claims to certainty. I would likewise ask the pragmatist to consider that the problem of certain knowledge does not confine itself to ivory tower epistemology, but extends its tentacles into pragmatic knowledge, as we shall see. There are serious implications and ramifications associated with the view that if a theory works, then it is a useful theory regardless of its "truth."

PREDICTING THE UNOBSERVED

The knowledge problem is most evident when we discuss our ability to predict that which has not yet been observed. Most people would say that they "know" the Sun will rise tomorrow. However, can we call this a certainty? It seems very likely, but it has also been predicted that the time will come (hopefully far in the future) when the Sun runs out of fuel, swells massively, and consumes the Earth. We don't know if this prediction is true, but it is consistent with our best understanding and we cannot rule it out. It is also possible that the Earth will explode due to some internal process with its molten core, which we had not anticipated. A massive comet that our telescopes have not observed may crash into the Earth and destroy the planet. These examples seem a bit extreme, but consider the 230,000 people who died in the tsunami in Sumatra in 2004, which resulted from an undersea megathrust earthquake that had not been anticipated. The most reasonable prediction on that day was that it would be an average day, like so many days before it, not that a massive wave was going to destroy many thousands of lives; tragically, such was the case.[3]

Another problem with the concept of knowledge is caused by the association of things. Humans are highly skilled at observing patterns of associations. Whenever there are dark clouds, sounds of thunder, and flashes of lightning, we consider it more likely to rain than if such things are not observed. The more people smoke

---

[3] Wikipedia. n.d. "2004 Indian Ocean earthquake and tsunami." https:// en.wikipedia.org/wiki/2004_Indian_Ocean_earthquake_and_tsunami

cigarettes, the more likely they seem to suffer from breathing problems, heart attacks, strokes, and lung cancer. Children who get measles vaccines appear to have higher rates of autism. However, while we are very good at observing patterns and associations, we often see patterns that are not there. More important, if the patterns are there, we often assert causal relationships between things (e.g., more people who smoke have lung cancer, therefore smoking causes lung cancer). However, we do not observe *causal relationships* between things; we only observe *sequences of events*. Thus the knowledge problem finds its place here as well, as we cannot observe causality and can only speculate as to causal relationships.[4] Our speculation is not limited to passive observation of association; indeed, all manner of tests can be conducted to get closer to the causality question (as will be explored in later chapters); however, at the end of the day we are limited to reasoning to the existence of causal relationships, which we cannot ourselves directly observe.

These problems are compounded when we are speculating about causal relationships between one thing and an additional entity that we cannot observe. If a person is found dead with a knife sticking out of his or her chest, an investigation is launched to identify the person (or persons) who murdered the victim; however, this action is based on an assumption that someone did murder the victim, and thus we are ascribing a cause to an unobserved source. When people have symptoms that resemble the flu, we ascribe the cause of their illness to a microscopic virus that we do not directly observe. Even the diagnostic test we employ to "confirm" influenza does not typically observe the microbe, but rather observes other effects of the microbe (e.g., antibodies in the patient). No human has ever observed a magnetic field directly; rather, the effects that we observe on magnetic metals cause us to evoke the concept of a magnetic field. In recent

---

[4] There is a rich and well-developed literature on issues of causality and what it means. Perhaps the most famous philosopher who pointed out that we don't observe causality was David Hume. Hume D. 1748. *An Enquiry Concerning Human Understanding.*

years, astrophysicists have postulated the existence of "dark matter" that makes up the majority of stuff in the universe and which causes the stars and planets to have their current locations and orbits, yet no one has directly observed dark matter. In these cases, the problem is not only that we are postulating a cause and effect relationship that we cannot observe between two entities, but that we also cannot directly observe one of the two entities posited in the causal relationship.

Certainly, things exist that we cannot directly observe. Denying this would be intellectually paralyzing, because we could only act on that which we could perceive. In such a case, I would have to assume that nothing existed behind walls that I could not see through. Our inability to observe things doesn't mean they don't exist, nor is it a problem to use theories that assume their existence if such theories help predict the natural world in a meaningful way. However, it is a problem in the context of knowledge of the unobserved entities and their associations. Evoking an unobservable entity that may explain observable things does not mean that the unobserved entity actually exists, any more than not perceiving it means it does not exist. We shall explore this more deeply in Chapter 2.

To fully explore the depth and manifestations of the knowledge problem, and how it really is a human problem, it is necessary to explore types of reasoning employed by humans, as the nature of human thinking leads to both the benefits and problems of human understanding. This is a separate topic from issues of human cognitive errors (e.g., the common sources of misperceptions or errors in reasoning); rather, it is an exploration of the limits of knowledge even in the context of correct perceptions and cognition.

INDUCTION AS A BASIS FOR THINKING

Experience and the ability to learn from such experience convey fundamental advantages to any creature that can modify its behavior based on past events. This is why memory is so important. As we catalog our different observations throughout life, we gain wisdom that can give us profound advantages over those who are less experienced or completely inexperienced. If you had to subject yourself to a surgical procedure, would you prefer a surgeon who had successfully performed the same operation hundreds of times or a doctor who had never done the operation even once? The second or third time you travel through a foreign country, travel through an airport, or even go to a restaurant, you have abilities that you didn't previously have. Basically, you "know the drill" – where the bathrooms are, what the different lines are for, what documents you need, and what the culture is like. Do you remember your first day of high school? For many of us it was a terrifying thing for a number of reasons, not the least of which was not knowing how to navigate an unfamiliar environment (forget for a moment that the madness of adolescence was clouding our feeble minds). However, as days and weeks went by, we became familiar with the place and the process, and were able to navigate a system and structure that we previously found confusing and intimidating.[5]

Induction is a natural form of human thinking that is practiced routinely and often unknowingly, and is required for everyday navigation of the world. It is basically the use of experience to predict events that one has not yet encountered. I distinctly remember an argument I had with my daughter in our kitchen one Saturday morning. She was 7 years old at the time and quite displeased with whatever it was I was telling her. She folded her arms across her chest, scrunched her face in frustration, and blurted out, "You can't predict the future!" My response was, "Of course I can. I predict that if I push this salt shaker off the counter, it will fall." Which I proceeded to do. She responded, "That's not what I meant. You can't *really* predict the future." She summarily dismissed my argument and stomped off in frustration. This incident illustrates a point that gives thinking animals with

---

[5] As was very much the case in my own experience, gaining the ability to navigate does not imply any manner of success or social acceptance; however, at the very least, I had a better idea of what humiliations to expect.

memories a profound advantage over other kinds of creatures. In fact, I had predicted the future and the prediction had held. It wasn't a stunning or unexpected prediction, and it was in a very limited context, but the fact remains that I had predicted the outcome of an event that had not yet occurred, and my prediction was spot on correct. I foresaw that the salt shaker would fall, as every previous salt shaker I had ever dropped had fallen; I had induced the prediction.

In more general terms, induction can be described as predicting the quality or behavior of the unobserved based on the observed. When you are only concerned with what has already been observed, that's not induction, it's description. In other words, if one were to restrict statements of knowledge to that which has already been experienced, the observations speak for themselves. I might simply state that every salt shaker I have dropped from my hand has fallen. It would actually be safer to state that I *perceived* every salt shaker I remember dropping to have fallen. If one restricts statements to the already observed, then one can make very clear statements about the perceived properties, but no predictions about the future are being made. Again, this is not induction but observation and only leads to encyclopedic information about things and situations already encountered. In this case, knowledge is no longer power, or at the very least a far less useful power, to the extent that power is the ability to predict and control – the ability to promote or prevent something.

Induction's immense power comes precisely from its ability to predict the future – that which has not yet occurred or been observed. However, this power comes with a tremendous vulnerability. The successful prediction of the falling salt shaker depended, as does all induction over time, on patterns in the future resembling patterns in the past. I have dropped a great many things in my life and almost all of them have fallen; indeed, every salt shaker I have ever dropped has fallen. So, it is easy to induce that when you drop things that are not otherwise supported, they fall. (The exception would be things that are less dense than air, e.g., helium balloons.) Yet just because things have behaved one way in the past does not necessarily mean they will continue to do so in the future – this assumption is the Achilles' heel of induction.

At first, this problem with induction seems a very common-sense sort of thing that doesn't set off any alarms. Everyone knows that things change, that things don't generally stay the same forever, and that there are times when past experience no longer applies. However, the gravity of this problem may be highly underestimated. A classic example of this problem with induction was put forth by Bertrand Russell, who described a chicken raised by a farmer. Every day of the chicken's life the farmer came out and fed the chicken. We'll assume that the chicken couldn't talk to the other chickens or to anyone else for that matter, and therefore the chicken's specific life constituted the entirety of its information. Hence, from the chicken's point of view, on every day that had ever existed, the farmer approached the chicken and gave it food. It would be a very reasonable induction for the chicken to predict that on the following day the farmer would once again give it food. Regrettably, when the next morning arrives, the farmer wrings the chicken's neck, plucks its feathers, and cooks it for supper – a tragic failure of induction to be sure – at least for the chicken.[6] The example of the chicken is highly applicable to human behavior. I have not yet died in a car accident; thus, I do not predict that I will die in a car accident today, and I feel comfortable driving – a mistake made by the more than 3,000 humans who die in car accidents each day worldwide. The assumption that the future will resemble the past is a highly useful assumption, but it is by no means certain. In some cases, it is almost inevitably false.

A practical example of falsely assuming that the future will resemble the past can be found with the advent of antibiotics. When

---

[6] The very same event may have been a failure of induction for the chicken and a successful induction for the farmer, who might frequently eat chickens he is raising. Of course, his perspective of the event that is being repeated is different – the raising of a chicken over time vs. day-to-day feedings.

penicillin was first used therapeutically in humans, it was observed that the administration of penicillin in patients infected with gonorrhea was uniformly efficacious in killing off the bacteria. One might be tempted to conclude a general principle – that penicillin kills gonorrhea. In fact, this became an accepted practical truth, and penicillin was listed by the medical community as the definitive treatment for gonorrhea. However, given the selective pressure of widespread penicillin use, some strains of gonorrhea acquired resistance to penicillin through evolutionary processes. Thus, whereas essentially 100% of gonorrhea was observed to be sensitive to penicillin in the past, such is not the case at the current time, a clear example of the fallibility of induction in being able to predict the future.

The problem of predicting future events by induction can be expanded to include the assumption that relevant modifiers of future situations will also resemble the past. In other words, the assumption that all things are equal – that one is always comparing apples to apples. I have a vivid recollection of the first time I gave my daughter a helium balloon (she was 9 months old at the time). When I handed it to her, she was extremely upset that the balloon "fell up" instead of falling down. It was an unpredicted event, because up to that point in her life 100% of everything she dropped had fallen down. Thus, it would have been very reasonable for her to predict that the balloon, like every other object, would in fact fall down. Induction failed in this case because the generalized rule did not happen to extend to this particular situation (i.e., helium balloons are not the same as other dropped objects). This problem with induction was not that the future didn't resemble the past, but rather that situational changes in the future didn't line up with the past. If I were to hold a salt shaker in my hand and then let it go while I was a passenger in the international space station, I would likely observe a very different result than in my kitchen on Earth. However, in the case of both the helium balloon and the space station, the future exactly resembled the past – as far as we know, helium balloons have thus far always floated in the atmosphere of Earth and salt shakers have always floated in outer space; the failure of my prediction was that I didn't understand how other circumstances and modifiers had changed.

The problem of background circumstances is ubiquitous and takes place in everyday interactions. We all know how frustrating it is to receive unsolicited advice from strangers that doesn't seem to apply to our situations. Most of us have seen a child having a meltdown in a public place and the parents (or other responsible adults) struggling to calm the child. For those of us who have been that struggling parent, it seems that onlookers have a variety of responses: sympathy, relief that it is not their problem, annoyance at being disturbed, and disapproval of the child, the parents, or both. In many cases the onlookers are critical of how the parents are handling the situation and, in some cases, can't resist giving "helpful" advice.

The problem with such advice is that every child is different, every parent is different, every child–parent dynamic is different, and there are all manner of specific modifiers that may affect a given situation (i.e., the child's pet might have died, the child might be on the autism spectrum, or the family may have different cultural norms or be facing some unusual stress, etc.). In most cases the person giving the advice has limited experience with a small number of children and yet feels comfortable generalizing his or her advice to this child, and maybe to all children. Of course, there are some generalities to human behavior, and certain advice may very well apply, but for obvious reasons, it may not. This is most acutely felt when one's parents offer advice (often unsolicited and typically obnoxious) about how their grandchildren are being raised, because their advice is no longer applicable (and in some cases no longer legal), coming from a generation when corporal punishment was not only allowed but encouraged, when car seats had not been invented, and when there was no problem with chain smoking in the nursery. Likewise, conversely, it is easy for children to criticize their parents' past behavior when held up to current norms, which were not in existence when the behavior in question was taking place. In all these cases generalizations about what "should be done" are being drawn that may not be valid, because the specifics upon which the generalizations are based may not apply to the situation in question. This represents a fundamental weakness in all experience-based predictions, or, in other words, a fundamental problem with induction – the situational specifics from which the experience was derived are different in the new instance.

Problems of induction are not restricted to generalizations and predictions about the future, but also extend to knowledge claims about unobserved entities in the present time. A classic example would be a naturalist who had observed a great many swans and noted that all of them were white. How confident can one be in the generalization that all swans are white, not only all swans of the future, but all other swans currently in existence that one has not observed? How many swans would you have to observe in order for the principle to be true that all swans are white? Would half of all swans be enough? How about nine-tenths? Regretfully, epistemologists have more or less reached the consensus that in order to be sure one must examine every swan. No matter how many white swans a biologist observes, all that has to happen is for one black swan to be seen. The moment that occurs, the conclusion "all swans are white" is rejected, regardless of the vast quantity of white swans that have been previously observed. In other words, the only way to eliminate this problem with induction is to limit one's statements to that which has already been observed, which as described earlier is no solution at all, because by doing this one is no longer inducing but describing. We are no longer generating knowledge of the unobserved based on principles derived from the observed, and thus the problem of induction has not been remedied. (On a somewhat comic note, arrival of Europeans in Australia resulted in the discovery of black swans,[7] thus demonstrating the point logicians had been making for some time.)

There have been a number of elegant defenses of induction, but at the end of the day they all appear to fail to solve the fundamental

---

[7] Of course, people indigenous to Australia had known of them for some time and probably had the view that all swans are black.

problems described previously. One such defense would be to state that all swans are white, and should one ever find a nonwhite bird that otherwise appeared to be a swan, then it would no longer be defined as a swan. This essentially makes the hypothesis nondisprovable through self-definition.[8] Another common defense of induction is that while it is not perfect, it has worked pretty well thus far, and hence can be assumed to continue working pretty well into the future. However, this is simply justifying induction with induction. In other words, it's not a problem with induction that things that worked in the past may not work in the future, because induction has worked many times in the past and will thus work in the future. This is equivalent to saying that I know the information I get from the Internet is true because of an article I read on the Internet saying all things on the Internet are true, or that I know the Bible is true because the Bible tells me so. A detailed cataloging of the different defenses of induction is well outside the scope of this book, but excellent discussions of this issue are available to the interested reader.[9]

No defense of induction has yet solved the problems I have described, but this isn't meant to imply that induction has not been an incredibly useful tool or that humans shouldn't continue to use induction. It is simply to illustrate that one cannot arrive at certain kinds of knowledge using induction. David Hume, who gave the most famous description of why induction is flawed, went so far as to say that not only are inductive predictions not certain, they are not logically supported at all.[10] In other words, it's not just that there is no certain basis to predict the Sun will rise tomorrow, but that there is no logical reason for this prediction whatsoever. Hume also expressed a certain gratitude that human behavior did not depend upon logical certainty and absolute predictions, as nothing would ever have been accomplished had we waited for such predictions before acting.

---

[8] This is an example of the "No True Scotsman" fallacy.

[9] Salmon WC. 1966. *The Foundations of Scientific Inference*. Pittsburgh, PA: University of Pittsburgh Press.

[10] Hume D. 1748. *An Enquiry Concerning Human Understanding*.

Humans cannot help but induce in all aspects of life, as it is one of the fundamental ways by which we navigate the world. A person who has complete amnesia, or who cannot form new memories and is thus deprived of the ability to induce due to lack of conscious memory and experience, is at a tremendous disadvantage in the world. Induction has thus far been superior to random guessing or untargeted trial and error; however, as stated previously, it is not a path to certain knowledge about unobserved things, and it can be (and will be) tragically wrong at times.

REJECTING POPULATIONS BASED ON SINGLE INSTANCES

The experience of life is distinct and particular to each of us. Forget for a moment that even when faced with the same experience, we may each perceive it differently; clearly, we each encounter a particular set of conditions and life events, and we each have a different interface with the world. While we may also incorporate the information of others through communication, we still have direct access to only a very small slice of the pie that is our world. Most of what is in existence (the universe) is simply not accessible to us, and we know little of even that to which we do have access. What percentage of people do you actually know in your hometown, on your street, or in your workplace? For the nearly 50% of Americans who live in large cities, it is likely that you know very few of the people in your general proximity and very little about them. Certainly, none of us has met a significant percentage of the approximately 7 billion people on Earth, seen a significant amount of the 197 million square miles of the Earth, encountered a significant number of the animals on Earth, etc. Yet, in order to use the power of induction to help us navigate the world, it is necessary that we make generalizations of some kind. Basing such generalizations on the small amount of data we have seems like a better guess than basing it on no data at all.

While we may be stuck doing our best to navigate the world with what we have, it is nevertheless a big problem to reject factual claims made about populations by using minuscule sample sizes. Nevertheless, this seems to be an enduring human trait. One reads that, on average, smoking increases the risk of getting lung cancer. This is a population-based argument. A group of smokers will have a rate of lung cancer that is 23 times more likely (for males) and 13 more likely (for females) than for similar populations who do not smoke.[11] However, this situation is often offered as the answer to the question: Does smoking cause lung cancer? When faced with such a statement, it is common to hear, "Well, you may say that smoking causes lung cancer, but my grandfather smoked four packs of unfiltered cigarettes for 35 years, and he never got lung cancer." This may have been the case for the grandfather, and that's a great thing for him, but it is irrelevant to the claim that smoking, on average, increases the risk of lung cancer. The claim was not that smoking causes lung cancer (i.e., if a person smokes, then he or she will get lung cancer) in the same way that removing someone's head causes death.[12] By definition, if smoking increases rates of lung cancer to anything less than 100%, then the population-based argument can be true even though some people will smoke their whole lives and never get lung cancer.[13]

Positive assertions of generalizations are no less based on minuscule data sets than are the rejections of assertions. One might go to two different restaurants and have a wonderful dining experience at one of them and a horrible dining experience at the other. Based on this experience, one rates the first restaurant as good and the second restaurant as horrible. However, the first restaurant may have gotten the wrong shipment of food that day, receiving excellent ingredients

---

[11] This statistic is in reference to small cell and non-small cell lung cancer (80%–90% of lung cancers). U.S. Department of Health and Human Services. "The Health Consequences of Smoking – 50 Years of Progress: A Report of the Surgeon General, 2014." www.surgeongeneral.gov/library/reports/50-years-of-progress/index.html

[12] At least so far, with the current limits of technology.

[13] There is also a percentage of people who never smoke and still get lung cancer. This compounds the problem of reconciling population claims with individual data due to a logical fallacy, because the claim that smoking increases rates of lung cancer in no way suggests that smoking is the only cause of lung cancer; it is not claimed that smoking is necessary for lung cancer to occur.

instead of the bargain basement, outdated food they normally buy to save money. In contrast, the second restaurant may have had both cooks and half their servers call in sick that day. The statements we tend to make are not that one particular meal was good at one place and poor at the other. Rather, we conclude one restaurant is good and the other bad, a generalized statement.

During the 2016 American presidential election, much regrettable rhetoric was passed around regarding whether or not individuals of the Muslim faith, or even of Middle Eastern background (regardless of faith), should be allowed into the country, or whether they should even be eligible to run for president if already citizens based on the assertion that people of the Muslim faith tend to be terrorists. As tragic as terrorist events have been in the Western world (and I use the Western world merely as a basis of comparison, not meaning to imply terrorism is any less tragic anywhere else), the perpetrators of these acts represent a very small number of individuals out of a world population of 1.6 billion Muslims (22% of all living humans on Earth). Surely, one cannot draw a meaningful generalization about 1.6 billion people based on the actions of a handful of individuals. If one were to look at Muslim-related terrorism in the United States, fewer than 20 individuals in recent years have engaged in terrorist acts out of 1.8 million Muslims in the country. This in no way rejects the observation that terrorist acts can be carried out by people of this group or that some extreme variation of ideology may drive the actions of these few individuals. However, this is a very small quantity of evidence to justify broader generalizations about Muslims. If anything, we can conclude that 99.9% of Muslims in the United States are not terrorists, the very opposite of what the rhetoric was suggesting. Moreover, this situation is a prime example of the availability heuristic (heuristics will be discussed in Chapter 4) combined with the base rate fallacy. When someone perpetrates a terrorist act, the media tells us the characteristics of that person. However, the media seldom (if ever) tells us the number of people with the same characteristics who don't carry out such acts.

This tendency to draw generalized knowledge from scant data may be the best we can do as individuals, as performing population-based studies is not a typical activity of humans; even if we were so inclined to engage in systematic study, most of us have neither the resources nor the ability to do so. However, the fact that individuals are doing the best they can doesn't mean their best is necessarily doing it well. Moreover, even when we do have access to the population data (e.g. with Muslims and terrorism), we are prone to ignore it. As will be discussed in more detail later in the book, it can be argued that the study of science has focused on (and analyzed) only a very few scientists and drawn general conclusions based on them. Moreover, by focusing on the scientists who have made the most progress (or at least are the most famous), those who study science bias themselves to the extreme of the population, potentially hobbling any ability to capture what scientists do in general (or as a group).

WHY PROBABILITY-BASED THINKING DOESN'T HELP WITH INDUCTION

A common approach to the problem of induction, which is often invoked in response to the previously stated concerns, is to state induced knowledge claims in probabilistic terms. This applies both to making statements about unobserved entities in the present and also across time. For example, if one had observed 99 ravens and all were black, then one might induce the statement that "all ravens are black." However, if the 100th raven observed was not black, we wouldn't throw up our hands in frustration at having no knowledge of ravens. Rather, one would simply modify the knowledge claim by saying that "99% of observed ravens are black." This maneuver is simply restating the data to modify a principle about all ravens. This can then be used to predict unobserved events from a probabilistic point of view; you can't tell what color the next raven you encounter will be, but you can say that 99% of the time it will be black and 1% of the time it will be nonblack – not with absolute certainty regarding the next raven, but with predictive power regarding a whole population and the relative likelihood of what color the next raven will be.

A probabilistic point of view can't predict an individual event, but there is no reason it can't make predictions about populations with great accuracy.

Although probability statements may bring comfort to some people, they fail to help much with the knowledge problem itself and with the issue of induction. The reason probability determinations do not help with the knowledge problem is that even if the probability statement is true with a capital "T", it cannot provide the ability to predict next events with certainty. While a probability statement can tell you the odds that the next raven will be black, the next raven can only be either black or nonblack.[14] Being able to state the likelihood that the next raven will be black is a type of prediction. Nevertheless, even if one has absolute knowledge of a population, it does not speak to specific cases, and thus one still cannot predict particular events. When most people talk to their doctor they don't want to know what their probability of getting cancer is; they want to know whether or not *they* themselves will get cancer.

Another problem with probability statements is that, much like simple induction itself, one can never rule out things changing in the future. After observing another 100 ravens, the 99% probability determination may change again, and in fact will change, unless 99 of the next ravens are black and one is nonblack. Thus, while the 99% probability determination may be better than random guessing, it is not knowledge about which we cannot be wrong. Let us retreat even further from our desire for absolute knowledge and stipulate that the more ravens we observe, the better and better our probability determination will become.[15] This seems a justifiable statement (often called the law of large numbers). This is just another way of saying that the closer we get to having observed every raven, the closer we

---

[14] This example uses categorical classifications and assumes that there are distinct colors as opposed to simply being a continuum of colors. Although it can be debated whether clean and distinct categories truly exist in nature, humans nevertheless tend to think in categorical terms, and there certainly does seem to be some basis (if not an absolute basis) for categories.

[15] This resembles a more Bayesian approach.

get to knowing the color that all ravens are.[16] This view and approach would be acceptable and would lead to certain knowledge (albeit probabilistic) if one could make the assumption that things are distributed around the universe in a uniform fashion. However, any clustering of any kind, either at a specific time or over time, destroys this principle, and there is no justification to support uniformity in the universe; indeed, there are ample data to the contrary. Let us back off even further and stipulate that we have made an absolutely correct probability assessment of the universe and that distribution of variability is not a problem. There is still no way of assessing whether the existing probability distributions in the universe will hold into the future, which brings us back full circle to the main problem of induction in the first place.

DEDUCTION AS A BASIS FOR THINKING

Deduction is a separate means of generating understanding and knowledge claims that suffers none of the problems of induction. This does not mean it doesn't have its own problem and limits, but at the very least they are different then the problems of induction. Aristotle's writings provide the earliest known Western codification of deduction, which he demonstrated in the form of syllogistic constructs. Aristotle defines a syllogism as "a discourse in which certain things having been supposed, [and] something different from the things supposed results of necessity because these things are so." This statement, although almost circular in its appearance, defines the traditional basis for deduction. A syllogism has premises (statements of fact) and a conclusion that appears to be "different" from either premise alone. For example, consider the following two premises.

Premise 1: All polar bears are white.
Premise 2: All bears at the North Pole are polar bears.

---

[16] It is important to point out that the law of large numbers indicates that it is the number of things you observe, not the percentage of things, that gives predictive power.

These two statements are presented as matters of fact known to the thinker. Based on these two premises, one can reach the following conclusion:

All bears at the North Pole are white.

Although no direct information is explicitly stated in any one premise regarding the color of bears at the North Pole, deduction based on the combined content of the premises leads to the conclusion regarding the color of bears at the North Pole. Hence, new understanding has been deduced by analyzing and combining the premises.

A more general form of the previous syllogism, but of the same construct, is as follows:

Premise 1: All As have the property B.
Premise 2: All Cs are As.
Conclusion: All Cs have the property B.

The tremendous strength of deductive reasoning is that if the premises are correct and the logic is valid, then the conclusions are certain to be true – not likely to be true, not probable, but incapable of being incorrect. This sounds an awful lot like the type of knowledge we're seeking when we talk about true knowledge. If correct premises and valid deduction lead to certain conclusions, then this sounds promising indeed. Of course, there are many fallacies in deductive reasoning, and, like any weapon of logic, if it is not wielded correctly the result can be incorrect conclusions, even from true premises.

Let's look at the following example:

Premise 1: All polar bears are white.
Premise 2: All polar bears live at the North Pole.
Conclusion: All bears at the North Pole are white.

The conclusion in this case is not a correct result of the premises. The reason is that while the second premise limited where polar bears can live (i.e., at the North Pole), this does not rule out that additional bears (nonpolar bears) also live at the North Pole. Hence, bears at the North Pole may consist of some polar bears and some brown bears. This possibility does not necessarily make the statement false, as it doesn't guarantee that brown bears will be at the North Pole; however, it doesn't rule it out and thus allows for the possibility that the conclusion is incorrect. In other words, the conclusion is not necessarily true and is thus a fallacy that doesn't lead to certain knowledge.

Like induction, deduction is a common tool of human reasoning, without which we wouldn't navigate the world as well as we do. While Aristotle may have first named and characterized deduction, it is not something that Aristotle invented. Rather, he described a process that, like induction, is a normal part of everyday human thinking. Deductive thought in humans can be found in children as young as preschoolers.[17] This is not to say that humans are perfect deducers; indeed, a whole body of studies has shown that we tend to deduce incorrectly, especially in certain circumstances.[18]

The correct application of formal logic is a highly complex and well-developed field, much of which is difficult to learn and certainly not intuitive. Nevertheless, like induction, deduction is a normal part of human thinking that we deploy as part of our navigation of the world. However, errors in deduction are also a normal human trait. Moreover, when we make such errors, we often feel as though we have reasoned our way to a correct conclusion, even though we have actually failed to do so. It is for this reason that logicians have invented specific ways to express logical statements, have defined different types of logic and the rules by which they work, and have made tremendous progress in such thinking. Indeed, much of mathematics can be described as a deductive language.

---

[17] Hawkins RD, Pea J, Glick J, Scribner S. 1984. "Merds That Laugh Don't Like Mushrooms: Evidence for Deductive Reasoning by Preschoolers." *Developmenal Psychology* **20**: 584–94.

[18] Evans, J St BT. 2017. "Belief Bias in Deductive Reasoning." In Rüdiger PF (Ed.). *Cognitive Illusions*. pp. 165–81. New York: Routledge.

While very powerful, deduction does not solve the knowledge problem. The first thing to note, which is a fundamental limit to deductive reasoning, is that it doesn't generate information about the unobserved; rather, it only reveals complexities that are already contained within the premises, but which may not be intuitively obvious until the deductive reasoning is fully carried out. In other words, no new information has been generated that wasn't already contained within the premises; nevertheless, without the syllogism, the fullest meaning of the stated facts could not be demonstrated and may not be appreciated. This seems to be a real limitation to deduction, as without the ability to make any predictions about the unobserved, our ability to predict or control is limited. However, this limitation can be overcome if the premises are universal, thus allowing the deduction of universal conclusions. In other words, consider premises that include the type of language of "every A is a B" or "no A is a B." Based upon such universal premises, one can deduce knowledge statements that apply to every instance of A, even instances that have not been experienced. Thus, one *is* deducing knowledge of the unobserved. This is one reason why deductivist thinkers tend to prefer premises of a universal type (e.g., all As are Bs), for without such universal premises the conclusions are not universal. If the conclusions are not universal, then one cannot make statements (with certainty) about unobserved things. If one has not achieved certainty of unobserved things, then one has not gained true knowledge (at least as we have defined it), and the knowledge problem remains unsolved.

If deduction can generate true knowledge so long as it uses premises of a universal nature, then where is the problem? The problem is in being able to determine a justifiable premise of a universal nature. For centuries, a number of notable philosophers have believed that humans have some inherent ability to recognize natural truths. However, in recent times, neurology's and cognitive psychology's understanding of human perception and thinking has advanced to the point that we now appreciate that humans can be pretty terrible at perceiving the world right in front of them, let alone coming up with universal statements of truth (this is explored in detail in later sections). If there is a single error in a premise upon which a deduced system of knowledge is built, then the whole system may come crumbling to the ground. If the premises are not certain, then the knowledge is not certain, no matter how good the reasoning. If there is no reliable source for certain premises, then deductive thinking cannot solve the knowledge problem.

Some of our greatest institutions have solved the premise problem by simply stating that a given premise is true. For example, the U.S. Declaration of Independence states: "We hold these truths to be self-evident, that all men are created equal, that they are endowed by their Creator with certain unalienable Rights, that among these are Life, Liberty, and the pursuit of Happiness." In other words, these truths are self-evident because we say so (so there!), and we will now build a system of beliefs based in part on this premise.[19] If the truths really were self-evident, then this might be okay, but what justification is there for such an assertion other than the authors stating that they hold them to be self-evident? – in other words, their opinion, and what is the justification that their opinions are correct? In the same fashion, many religions are based on an irrefutable premise that a given god or gods exist. Likewise, many systems of belief, even without a formal deity, state the presence of a force, or energy, or structure in the universe. Such premises of gods or forces are certainly not without evidence; indeed, evidence of the divine can be obtained through an experience of god, through observable consequences that would come about in the event that there was a god, or even through revelation. One can feel universal sources through a spiritual experience or perceive the effects of such forces in the world.

One might then argue that there is no knowledge problem for philosophies that invoke self-evident premises or for religions that consider the experience of a god or revelation of the divine as sources

---

[19] No claim is being made here that the American system of government was deduced, just that it makes claims of self-evident and universal premises.

of unequivocal truth. However, as will be explored in detail later, it should be noted that such systems do not typically deduce an entire system of belief, at least not in any formal sense of deduction, from the premises that are stated, and thus it is a kind of an apple and orange situation. Moreover, while feeling or perceiving something can be a compelling force in persuading an individual that the thing exists, perceptions and feelings are highly prone to error and misinterpretation, and thus do not provide a justification of knowledge that stands up to analytic thinking. This is not to say that such justification is not sufficient for theology or spiritual systems of belief, but it is clear that such justification is fallible. How many religions have existed in human history, many of which believed with certainty that they were the one true way? For this to be true, all of them, except one, must be wrong, and it is not clear that any of them need be right. Thus, theological revelation appears to be able to get things wrong. Hence, while religions typically state things in certain terms, and may lead to certain belief, they do not lead to certain knowledge. The making of claims with certainty and explanations of everything will be explored later as one of the criteria by which one can demarcate some categories of nonscience from science.

If we stipulate that humans have no access to fundamental premises, or first conditions, through either revelation or innate knowledge of such premises, then how is one to use deduction? If the premises are not certain to be true, then no matter how valid the deductive reasoning, the results are not certain to be true, which undermines the whole deductive program for generating knowledge. One might point to Euclid, who stated certain premises and was then able to deduce a complex geometry that was very useful in describing the natural world. Likewise, Sir Isaac Newton stated certain premises (laws of motion) from which he deduced a system of mechanics that could describe and predict motions of the planets with great accuracy and how forces worked on bodies in general. Isn't the amazing predictive capacity of these systems a validation of the correctness of their premises? Regrettably, as we shall explore later, such is not the case. Of note, given modern theories of special relativity and the curved nature of time-space, both Newton's and Euclid's systems are considered to be profound intellectual achievements of great theoretical and practical value, but ultimately, these systems are not entirely correct due to the premises being not entirely correct.

At the end of the day, there is no clear way around the major problem of deductive knowledge. In order to have any ability to predict unobserved nature, deduction must make statements that are universal. Due to the problems of induction, universal statements based on experience cannot be justified, and no other source of universal premises seems supportable.

Although both induction and deduction have the problems described, in real life, one nevertheless uses induction and deduction (or at least reasoning that resembles deduction) together to navigate the world. Induction provides the justification for premises based on experience (albeit an imperfect justification). Deductive reasoning helps reason forward from the induced premises to generate all manner of new understanding of association within the induced premises. Hence the combination of induction and deduction certainly leads to new ideas that would have come from neither alone, but fails to solve the problems of either. In aggregate, the knowledge problem is solved by neither induction, deduction, nor a combination of the two.

THE UTILITY OF UNCERTAIN CONCLUSIONS: IS THE KNOWLEDGE PROBLEM REALLY A PROBLEM?

It seems that a solution to the knowledge problem is not likely to be forthcoming. However, how much of an impediment is this? It brings us to the question: What makes useful knowledge, and does useful knowledge have to be universally certain to be meaningful? Many thinkers have adopted a pragmatist school of thought that has placed value on scientific theories if the theories work in the real world. If a theory predicts the natural world, then it is a useful theory, regardless of whether or not it is ultimately true. Knowledge may be flawed, to be sure, and it may not result in any kind of absolute, objective truth.

However, it is hard to ignore the explosion in science and technology that has transformed the world over the last four centuries. Most of this transformation was carried out using theories that were not only uncertain (as all scientific theories are) but which are now believed to have been disproven. While "wrong" they were nevertheless very useful theories. Whether or not the progress of science and technology is good, bad, or amoral, the fact remains that generation of imperfect, uncertain, and ultimately flawed understanding has had a very real effect on the lives of untold millions. Despite missteps and errors, the scientific process on the whole has been fruitful. Given the problems of induction, we cannot assume science will continue to work with any certainty, but it does not appear to have stopped working yet; it appears as though uncertain theories, albeit imperfect, can be pretty useful.

Of immense importance is the understanding that induction and deduction are tools in the toolbox of thinking, but that these are not in and of themselves methods of modern research. To be sure, there are modern inductivists (e.g., botanists in the rainforest cataloging new species of plants or those sequencing every bit of DNA they can get their hands on to generate encyclopedic databases) and modern deductivists (e.g., theoretical mathematicians). However, the important message here is the recognition that induction and deduction are parts of normal human thinking. While they are employed by scientists, they are also employed by basically everyone else. Thus, the weaknesses of induction and deduction are weaknesses of science and nonscience alike. Because they are ubiquitous, the simple use of induction, deduction, or both in combination cannot be a criterion to distinguish science from nonscience. Yet, while common to both science and nonscience, induction and deduction are nevertheless integral and essential to the scientific method, and therefore are essential trees to be understood as we continue to develop, in upcoming chapters, a view of the forest that will help distinguish science from nonscience.

2 Adding More Building Blocks of Human Reasoning to the Knowledge Problem

> Not the smallest advance can be made in knowledge beyond the stage of vacant staring, without making an abduction at every step.
>
> – Charles Sanders Peirce

RETRODUCTION AS AN ADDITIONAL TOOL OF REASONING

Induction and deduction, as discussed in the previous chapter, have received a great deal of attention from multiple quarters. In the nineteenth century, the philosophers William Whewell and Charles Sanders Peirce focused on retroduction as a distinct mode of reasoning. Retroduction had been recognized by Aristotle as a separate entity with specific properties; however, it wasn't until Whewell and Peirce that a strong distinction between retroduction and induction was emphasized.[1] Retroduction is an essential part of human reasoning, without which ideas of causal relationships essentially could not expand, as induction and deduction can only get one so far. Indeed, Peirce (who, it can be argued, was most instrumental in recognizing the role of retroduction in science) described this mode of reasoning as "the only logical operation which introduces any new idea and commented: "[N]ot the smallest advance can be made in knowledge beyond the stage of vacant staring, without making an abduction [retroduction] at every step.[2]"

---

[1] Unfortunately, Whewell called retroduction "induction," which confused it with earlier meanings of the word – as such, we will avoid that term here.

[2] Charles Sanders Peirce referred to retroduction as "abduction" in his writings.

44

What, then, is retroduction? It consists of the process by which one generates ideas regarding the causes of things already observed; in Peirce's words, retroduction is "the process of forming explanatory hypotheses." In other words, retroduction enters our thinking with regards to the association between observed effects and the things we speculate caused those effects. Retroductions are to be found everywhere and in every walk of life, as a ubiquitous part of normal human thinking. One awakens one morning to see snow on the ground that was not present the night before, and retroduces that it must have snowed during the night. One arrives home to see a spouse's car in the driveway and retroduces that one's spouse has already arrived at home. One receives an email from the address of a friend and retroduces that the friend sent the email. Based on evidence, we are guessing at a cause of the effects we observe.

One could argue that this is just experience-based thinking – the simple carrying out of enumerative induction – for example, all of the previous times when there was snow on the ground it had snowed, previously when one's spouse's car was in the driveway, the spouse was in the house, and when previous emails were sent from a certain address, it was a particular friend who had sent them. However, retroduction is quite distinct from induction. It is not observing something and predicting a general principle from what is being observed; it is not a "more of the same" conclusion. Rather, it is suggesting a *previous* (hence "retro-") entity that led to the observed outcome. With retroduction, one is typically positing something that occurred in the past to explain the cause of an experience, not predicting observations that have not yet been experienced. Moreover, retroduced causes can be entirely novel things that have never been observed (e.g., positing that an invisible evil demon is the cause of an outbreak of an illness). Hence, retroduction is distinct from induction. Whereas induction may lead to generalized conclusions and predictions of as of yet unexperienced things, retroduction makes a retrograde guess at the causes of an already observed phenomenon. Importantly, retroduction need not be limited to temporal and causal entities but can also apply to laws and principles that explain without invoking cause; however, for the current discussion we will focus on retroduction of causal entities as our main example.

Although retroductions may posit as of yet unknown causes of observed effects, good retroductions don't just randomly guess. Rather, the retroduced cause or causes should only be entities from which the observed effects would follow, or at the very least be consistent. An example might be taken from cancer epidemiology. Let us say there are an unusually high number of childhood leukemia cases in a small Arkansas town; we will assume this observation to be accurate and correct. One induces the generalization that children living in this particular town are at an increased risk of getting cancer. The process of retroduction would begin with the generation of a hypothesis capable of explaining the induced generalization (i.e., increased likelihood of leukemia).

For example, suppose one retroduced the presence of a carcinogen in the drinking water of the Arkansas town and that this particular carcinogen caused higher rates of leukemia in children. Let us also stipulate that the children in this town frequently consumed the drinking water. All other things being equal, then one can deduce what was observed – that there would be higher rates of leukemia in the children in this town. The term "all other things being equal" is a big logical lift, and it is difficult (if not impossible) to ever justify such a condition in the real world, as we shall explore later. Nevertheless, scientists often think in this way, and even if a formal deduction is not possible, one can at least infer the observed outcome based on this retroduction. It is not an acceptable retroduction to guess at a cause that wouldn't lead to the observed effect. In other words, if we accept that consuming candy has no link to cancer, but retroduce that the higher rates of cancer in the town are due to eating more candy, this retroduction is not consistent with accepted evidence. Even if children eat more candy, it would not lead to that which we are trying to explain. This is not to say that people don't make retroductions that wouldn't lead to what is being explained – but at the very least, these

are bad retroductions. For the retroduced hypothesis to be scientif-
cially useful, it must at least have the potential to lead to the known
effect. To be scientifically useful, it must also lead to other (as of yet
unobserved) effects, as we shall explore later in the text.

Retroduction, then, is a third and separate mode of reasoning,
and by adding it to induction and deduction an integrated model of
scientific reasoning can begin to emerge. However, before such a
model is synthesized, the problems of retroduction need to be
explored in more detail. Just as retroduction is distinct from induction
and deduction, its advantages and problems are also distinct.
A specific problem with retroduction is called the "fallacy of affirming
the consequent" and neccessitates some exploration.

THE FALLACY OF AFFIRMING THE CONSEQUENT

Consider this statement: if A, then B. In other words, if A is the cause,
then B must be the result. As an example, one can say that if an
individual falls off the roof of a 30-story building and lands on concrete
(all other things being equal[3]), the person will be injured. If Bill unequivocally falls off a 30-story building, you can conclude with certainty that
Bill will be injured.[4] If A, then B; A occurs, therefore B must occur.

However, one cannot logically go in the other direction; in other
words, given the statement: If A, then B, one cannot conclude that if
B occurs then A must also have occurred. Why is this so? If one finds
Bill lying on the sidewalk injured, why can't you conclude with
certainty that Bill fell off the 30-story building alongside the sidewalk?
The reason is that all other kinds of causes may have resulted in Bill's
injury (e.g., being hit by a car, having fallen out of a plane, being
clumsy and falling down). In other words, as illustrated in Figure 2.1,

---

[3] In this case, "all other things being equal" means without introducing any other
modifiers (e.g., the person doesn't have a parachute, the building is on Earth, the
rules of gravity apply)

[4] It is not being claimed that there has never been a report of someone falling off a 30-
story building and not being injured, but as an example, we can accept that all will at
least be injured somehow – if only a scratch or bruise.

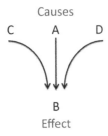

Causes

C   A   D

B

Effect

FIGURE 2.1 Cause, effect, and affirming the consequent

consider that A, C, or D each cause B. If one knows that C or A or D occurred, then one can conclude with certainty that B occurred. However, knowing that B occurred, one cannot conclude with certainty that A occurred, as B might have been caused by C or D (or even by some as of yet unknown cause not in the figure).

Why is this fundamental issue so important? Because a great deal of normal human thinking and interacting with the world is precisely the process of observing B and positing A, which is basically the process of retroduction.

In our personal lives, our professional lives, and in local and world events, we are constantly observing effects and retroducing causes. We observe that the Earth is getting warmer, and people propose different causes. We observe that one country invades another, and we speculate as to the motivation. Each day the stock market goes up or down, and all manner of financial punditry puts forth multiple theories about what caused the change.

At a more basic level, someone wins the lottery and everyone begins to seek a reason, such as where the lucky person purchased the ticket or what clothes the person was wearing when he or she did so. People get sick and we have no explanation, so we begin to suggest causes (e.g., toxins in the water, or radio towers, or plastics, or vaccines). The classic murder mystery is still another example – a dead body is found, and the detectives guess at suspects and try to figure out who murdered the victim.

In each of these cases, many different people simultaneously posit various theories. Although we don't normally use the word, in all of these cases people are retroducing hypotheses to explain observed effects, which is a common process. Moreover, in doing so, people are evoking the fallacy of affirming the consequent, and this is why the speculation goes on so long. Affirming the consequent is an inexorable logical defect in the process of retroduction itself, and thinking alone cannot remedy this defect. This doesn't mean retroduction may not get the right answer; it just means we can never be certain that it has. It is for this reason that it has been questioned whether retroduction is even a form of logic at all; however, it is certainly a form of thinking, and its utility seems clear (if not at all certain).

So, in the case of the town in Arkansas, how many different hypotheses are equally consistent with the data about high rates of cancer? The answer is that there are an infinite number, limited only by the imagination of the thinker.

1. There is a toxin present in the town that causes cancer.
2. The townspeople are infected with a virus that causes cancer.
3. There is a cancer-causing genetic mutation present at a higher rate than in the general population in the families that live in the town.
4. There is a hole in the ozone layer over this city, resulting in more cancer-causing ultraviolet sunlight affecting the citizens.
5. There are high voltage electrical lines in this city that cause cancer.
6. Large deposits of magnetic rock near the city cause cancer.
7. The combination of toxins and a magnetic field are causing cancer.
8. There is a toxin in the fish living in a nearby lake, and eating this fish causes cancer.
9. Secret government radiation experiments are being carried out on the town's children.
10. Space aliens are abducting the kids and implanting them with cancer-causing probes.
11. Past immoral behavior by the kids' parents has caused bad karma and cancer.
12. God is punishing the town.

Here we see the fundamental problem with retroduction and with all hypothesis-based thinking. For any given observation, there are an infinite number of hypotheses, each of which would equally result in what we have observed. This is not to say all possible hypotheses are equally consistent with the data – many things just don't lead to what we have observed and are not considered. But, of the hypotheses that would lead to the observed data, there is no limit. Your neighbor may be a trickster who put snow on the ground and thus it didn't snow last night; someone else may have driven your spouse's car to your house and thus your spouse may not be there; and a stranger may have hacked your friend's email account and sent you a message.

Another way to understand the problem of affirming the antecedent is through mathematical representation. For example, one might be given the mathematical equation $x + y = z$. If one is told that $x = 10$ and $y = 20$, then one can solve that $z = 30$, and this is the only answer that fulfils the equation.[5]

In contrast, due to its basis in observation of the natural world, retroduction starts with the outcome. In other words, we know the answer because we have observed it in the world around us, and we are seeking to identify the cause or causes. Thus, the retroduction process starts with the equation $x + y = 30$.

Well, this is easy to solve. For example, $10 + 20 = 30$. This is a perfectly valid solution to the equation, with $x = 10$ and $y = 20$, which lead to a z value of 30. However, $5 + 25 = 30$ is an equally valid solution, as is $1 + 29 = 30$. Allowing negative numbers and decimals, it is easy to see that there are an infinite number of mathematical solutions to this problem, all of which are equally correct.

This is the first major problem with retroduction, and, in fact, with essentially all thinking about causality – for every explanation for the observed phenomenon, there are an infinite number of alternative explanations that are equally consistent with the data at hand. This doesn't mean that you can't rule out invalid retroductions (e.g., $x = 20$, $y = 20$, in which case $x + y$ does not equal 30); however, after

---

[5] This is assuming base 10, normal rules of mathematics, etc.

excluding invalid retroductions, one is still left with an infinite number of valid retroductions. In using retroduction, one has no way of telling which of all the valid retroductions are (or are not) actual causes. One cannot assess the likelihood that a posited entity is causal by reasoning alone.

THE INTEGRATION OF INDUCTION, DEDUCTION, AND RETRODUCTION INTO A SYSTEM: HYPOTHETICO-DEDUCTIVE THINKING

With the concepts of induction, deduction, and retroduction now defined, one can synthesize them together into a system of thinking called Hypothetico-Deductivism (HD; Figure 2.2). HD has been used to describe a process of thinking that has been presented as a model of how science works. This process consists of observing facts about the natural world and retroducing hypotheses (regarding causes) from which one can deduce that which has been observed, as described earlier. In this case, the prefix "hypothetico-" represents retroduction, as retroduction leads to hypotheses regarding the causes of observed effects. The word "deductivism" means that one must be able to deduce the observation from the retroduced cause. It should be noted that many hypotheses have a probabilistic or statistical relationship between the retroduced cause and the effect that follows. In such cases, since the outcome only has an increased likelihood of occurring (e.g., the cancer example discussed previously), it is not certain that

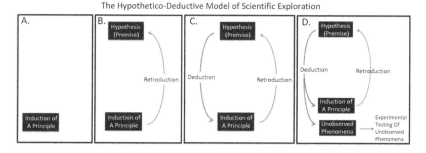

FIGURE 2.2 The hypothetico-deductive model of scientific exploration

the cancer outcome will occur – as such, this is not deduction in the traditional sense. Nevertheless, the increased rates are predictable, and for the rest of the text we will use HD to include these types of examples.

HD then takes an additional step, which has emerged as an attempt to address the fallacy of affirming the consequent; in other words, to help narrow down the number of possible causes of an observed effect. This step is to use deduction (or statistical inference in the case of probabilistic hypotheses) to make additional predictions, which have not yet been tested (e.g., no observation or attempt at observation has yet been made; Figure 2.2D). The importance of this additional point cannot be emphasized enough, as it provides a response to the problem of affirming the antecedent (albeit an imperfect response). If new effects that can be predicted from a retroduced cause do not occur, then that cause is no longer a valid retroduction, as its presence would not lead to observed outcomes. While all the retroduced hypotheses may predict the initial observation or observations, different hypotheses ultimately lead to at least some different predictions, by which the hypotheses can be whittled down.[6] Thus, by identifying additional predictions and testing them, the validity of retroduced ideas can be assessed. Although philosophers and scientists have used HD to characterize scientific thinking, there is nothing uniquely scientific about HD thinking. As is the case with induction, deduction, and retroduction, HD thinking is found in everyday human thinking and problem solving.

Let's look at an example of HD thinking in everyday life. One cold morning you get into your car, turn the key, and the engine doesn't turn over. You have observed that your car doesn't start. You retroduce that the cause is a dead battery. You have just used the first part of the HD method – you have retroduced a hypothesis to explain why your car wouldn't start.

---

[6] Arguably, if two hypotheses do not lead to any differences in deducible outcomes, then they are functionally indistinguishable as different hypotheses.

Premise 1: Cars need charged batteries to start.

Premise 2: My car won't start.

Retroduction: Therefore, it would be reasonable to hypothesize that my battery is dead.

Thus, a cause/effect thought construct has been generated that is logically coherent, which is to say that given the background information you have, the hypothesis (i.e., my battery is dead) predicts the effect (i.e., my car won't start).[7]

In a more perfect world, without logical fallacies, you would have the correct answer. However, due to the fallacy of affirming the consequent, you have a serious problem. One can retroduce multiple other hypotheses, all of which will also predict that the car won't start. For example, the ignition switch or the starter motor may be broken, the fuse system may be fried, a vandal may have popped the hood and stripped out the electrical system, or there may be no engine in the car at all, since you forgot you removed it yesterday. Indeed, as with the example of cancer in the small Arkansas town, there are an infinite number of hypotheses explaining why your engine won't start, once you get creative enough. Space aliens are beaming a signal into your car to prevent it from starting. Demons have taken over your car. Someone has substituted a car identical to yours that doesn't start.

To further illustrate this point, the following retroduction is a coherent explanation of why your car is not starting.

Retroduction: The spirit of Elvis Presley has become very angry about global warming and has thus extended his essence into all internal combustion engines in the world, preventing them from starting due to the all-powerful influence of the King of Rock and Roll.[8]

---

[7] It is very important to note that this thought process involved the use of essential background information regarding how cars work.

[8] Of course, this example must include background premises, such as your car runs on an internal combustion engine. In addition, this retroduction, while coherent with the observation that your car doesn't start, is not an acceptable scientific hypothesis for reasons that will be explored later. However, it is used here to illustrated that many odd or magical hypotheses, while not scientific, nevertheless have a

Retroduction is a common and ubiquitous part of human thinking; at the same time, affirming the consequent is a problem with all retroduction, since there are infinite causes that could predict any outcome. Indeed, by reasoning alone, there is no logical way to assign the dead battery hypothesis any more truth than the Elvis Presley hypothesis. Induction may seem to help, since you have had dead batteries before but never a problem with Elvis (at least not that you know of), but as explained in the section on induction, while a dead battery may indeed turn out to be the problem, this position is not justifiable by logic.

It is at this point that you use the additional predictions of your hypotheses, combined with observation (both passive and active), to narrow down your hypotheses. By testing new predictions, one can begin to substantially refine the list of retroduced explanations that remain valid. You note that in addition to the engine not starting, the lights on your dashboard did not come on when you turned the key. This observation would be predicted by the battery being dead, the fuses being out, or the wiring having been stripped by a vandal; however, this would not be predicted by the starter motor being broken.

It follows from the dead battery hypothesis that replacing the battery would allow the car to start (this does not follow from the other hypotheses). So, if you then replace the battery with a new one, turn the ignition, and the car starts, you have now, in essence, performed an interventional experiment (as opposed to a passive observation). You have replaced the battery and the car started. We will assume that this is the only change you have made, indicating that the reason the car didn't start was due to a faulty battery, supporting that hypothesis and rejecting the others. You have just used the HD method to predict and control an aspect of the natural world.

retroductive component. Retroduction is found in all manner of common human thinking.

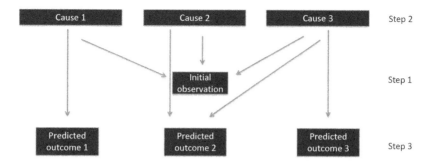

FIGURE 2.3 Interactions of different webs of belief

The logical basis for this thinking is depicted in Figure 2.3. An initial observation is made (Step 1). In an attempt to explain the initial observation, multiple causes are retroduced (Step 2), each of which would lead to the prediction of the initial observation. From these retroduced causes, one can predict additional outcomes (Step 3), each of which is predicted by some, but not other causes. One then investigates the predicted outcomes, either through observation or interventional experimentation, to assess if the outcomes are observed or not. For example, if one observes predicted outcome 2, this would support Causes 2 and 3 but exclude Cause 1, whereas predicted outcomes 1 or 3 would support only Causes 1 or 3, respectively, and exclude Cause 2.

As we shall see subsequently, the reality of the situation is much more complicated, but this is the basic framework for how the HD method works. In our car example, the initial observation was that the car didn't start. Of the multiple causes, some predicted that the dashboard lights would work, whereas others predicted they wouldn't; thus, observing the dashboard lights helped to narrow the field of retroduced hypotheses. Of the hypotheses tested, only the hypothesis that the battery was dead predicted that replacing the battery would solve the problem.

The fixing of the car by replacing the battery essentially ruled out all other stated hypotheses and supported the dead battery hypothesis. Many people (including many scientists) would say that you proved that the dead battery was the problem; regrettably, such is not the case. Why not? Although the example seems very straightforward, one can always retroduce another hypothesis that explains the observed phenomena. Perhaps a bunch of gunk had accumulated on the leads of the old battery, preventing an electrical connection to the starter motor. In this case replacing the battery made the car start, but not because the old battery was dead, rather because you had inadvertently cleaned the leads through the friction caused by changing the battery.

This illustrates HD's problem-solving abilities (after all, the car now starts), but also its inability to achieve certainty: we can't be entirely sure why.[9]

LYING YOURSELF INTO A CORNER: A CLASSIC EVERYDAY VERSION OF HD THINKING

Consider the time when you spent the week at the beach relaxing in the summer sunshine instead of going to work. On the first day, upset by your absence, your boss calls you on your cell phone. Faking your best "sick voice," you tell her you awakened with a fever and chills that morning, that you have the flu, and that you are going to miss work until you recover. You basically provided your boss with a cause that predicted the observation that you were not at work. Your explanation was provided after the effect (you not being at work), but it described a cause that preceded you not being at work (waking

---

[9] Science has often been likened to problem solving, not unlike fixing the car. However, practicing scientists often don't stop once they have solved the problem. For the mechanic, the car now works properly, and so the task at hand is done. Scientists also spend time testing other deducible predictions of their working hypotheses to see if the predictions hold, even if they have no obvious practical utility. Indeed, scientists often do things against a practical goal (e.g., they might go back and purposefully "re-break" the car in different ways to see what "would have happened if"). Because greater understanding often has unanticipated practical utility, this still fits with practical goals. However, its motivation is more focused on understanding a system than on specifically solving a practical problem like getting the car to run.

with a fever and chills). In other words, one could predict the outcome that you would not be at work if given the hypothesis that you were too sick to get out of bed. This is the equivalent of retroducing a hypothesis consistent with observations.

However, unbeknownst to you, your boss called you at home several times later that week (just to see how you were feeling), and of course you didn't answer any of the phone calls because you were at the beach. This was a surprise to your boss, because one could also predict (all other things being equal) that if you were sick at home you should answer your phone.[10] Your boss was worried that maybe you were unable to answer the phone because you might be incapacitated and an ambulance should be called. However, she also considered that you might be out seeing your doctor or that you might not actually be sick at home and were taking unauthorized time off (another example of retroducing multiple hypotheses, each of which predicts that you wouldn't answer the phone). Your boss decided to just wait (rather than conducting an investigation, like calling your cell phone again, going to your house or calling 911). When you return to work, she asks why you didn't pick up the phone. Worried about your lie being discovered, you say that you had the landline disconnected and are only using a cell phone. You have modified the conditions of the hypothesis by adding an auxiliary statement of no landline, so that once again the outcome is predictable from the hypothesis along with the auxiliary statements (*I was too sick to come to work, and so I stayed home, where I don't have a phone line hooked up*).

Regrettably, while talking to you, your boss notices that you have a new tan! Your tan was certainly not predictable from the hypothesis; to the contrary, one would assume you were inside and very much away from the Sun. In a maneuver born of slowly increasing desperation, you mention to your boss that you have recently purchased a tanning bed for your home, and that you were in it for a

---

[10] Of note, this phone is assumed to be a landline and not a cell phone. For any readers born after 2010, the definition of landline can be looked up on Wikipedia.

half hour a day while you were sick because it made you feel warm. Once again, one can predict the observations from the hypothesis – you were absent from work, weren't answering your phone, and are now tan from the new modified hypothesis (*I was too sick to come to work, so I stayed home, where I don't have a phone line hooked up, but where I do have a tanning bed*). Encouraged by your description of how good the tanning bed felt, your boss asks if she can come over to your house this afternoon and use your tanning bed to get ready for her upcoming trip to the Caribbean. From your new explanation, your boss can predict that if she walks into your home she will encounter a tanning bed. In reality, if your boss walked into your house she wouldn't encounter a tanning bed (and she might also see a telephone hooked up to a landline), at which point you'd have to come up with another lie to explain the absence of the tanning bed and the presence of the phone – either that, or concede that you were lying all along.

Lying yourself into a corner, which most of us have experienced at some point in our lives, is essentially the HD process. It occurs in our personal and professional relationships and is a large basis for how police and prosecuting attorneys pursue investigations of the accused. An accused person provides the alibi that he was hundreds of miles away from where a crime occurred. The police gather all the information they can that can be predicted from the alibi, such as witnesses, phone records, credit card charges, security camera recordings, etc. to see if it all lines up. If the evidence is not consistent with your alibi, then the police conclude you were lying to them and you remain a suspect. Indeed, just as scientists often speak of ruling out a hypothesis, so do law enforcement officials speak of ruling out (or eliminating) a suspect. In fact, many different types of investigations can be carried out by this process.

Anyone who is a parent has lied themselves into a corner with their kids, and it is easy to see even young children participating in the HD process. As an example, in Western society, there is strong evidence that Santa Claus exists. Presents appear under the tree, and the Santa myth is presented as the cause of the observed effect. But then the child asks: How does Santa make all those toys, how does he get into our house, and how does he get to every home in the world? Answers are supplied; he has an army of elves making his toys, he comes through the chimney, he has a flying sleigh pulled by magic reindeer. But with these new premises the child can make new predictions, leading to subsequent questions: How do they feed all the elves, how does he get into houses and apartments without chimneys, and don't the reindeer get too hot when they fly to the equator? If you are part of this culture and have children then you know the rest of this drill.

USING HYPOTHETICO-DEDUCTIVE THINKING TO DIFFERENTIATE BETWEEN SCIENCE AND NONSCIENCE

HD has been proposed as a model of how science works and that it can be used as a criterion to demarcate that which is science and that which is not science. However, there are problems with this argument; these problems fall into several different camps.

The first problem is that HD fails to provide any true "knowledge" consistent with the overall knowledge problem described previously.

In the words of John Stuart Mill:

> Most thinkers of any degree of sobriety allow that an hypothesis...is not to be received as probably true because it accounts for all the known phenomena, since this is a condition sometimes fulfilled tolerably well by two conflicting hypotheses...while there are probably a thousand more which are equally possible, but which, for want of anything analogous in our experience, our minds are unfitted to conceive.
>
> *[1867[1900], 328][11]*

---

[11] Mill JS. 1900. *A System of Logic, Ratiocinative and Inductive.* New York and London: Harper & Brother Publishers, p. 328. The issue being raised here by Mill is also one related to underdetermination of theories, which will be discussed later. Included in this issue is how can one ever systematically test and/or compare

This objection is only a concern if one holds that science is defined as being fact and thus cannot tolerate lack of truth. But this objection doesn't hold water once we concede (as I think we must) that science is certainly flawed, given all the mistakes science has (and continues) to make. Since science is imperfect and fallible, then the fallibility of HD does not prevent it from serving as a description of scientific practice. Along these lines, if science is meant to be a practical way to increase our ability to predict and control the natural world, and does not carry the ambition of finding underlying "truths" of nature, then there is really no objection at all.

The second problem is a much more serious concern. As shown by the example of the nonstarting car, HD thinking is common human thinking in everyday life carried out by nonscientists; because it is part of everyday thinking and is found everywhere, this should eliminate HD as a sufficient condition of doing science (unless we say that all of us are doing science all the time, which is neither the case nor helpful). This is not to say the mechanic is not clever or useful – I don't know many scientists who can fix a car, but we don't typically call mechanics scientists. Perhaps more troubling, there is tremendous evidence that practicing scientists (even world famous scientists with great accomplishments) carry out research and discovery in a fashion that often has little resemblance to HD at the time it is being carried out. Indeed, science is often reported and represented, after the fact, as a process based on HD, but such is typically not the way it is really carried out (this is explored in detail in Chapter 10).

Thus, HD thinking is not unique to scientists and may not even be the process by which science is done in real time. Hence, it seems impossible to justify a position that HD thinking is a defining factor of science. However, this does not mean HD thinking isn't a critical component of scientific practice. In later chapters, the notion will be put forward that whereas engaging in HD itself does not distinguish

hypotheses if there are an infinite number of hypotheses equally consistent with our data and observations.

science from nonscience, the way in which HD is ultimately used by scientists does in fact contribute to the demarcation of science from nonscience. For the current chapter, gaining an integrated view of how induction, deduction, and retroduction fit together as constituents of a method of thinking, which appears to be common to human cognition, is an important basis in moving forward to build a definition of science.

THE PHANTOM NATURE OF SCIENTIFIC ENTITIES

As early as 1667 and throughout the 1700s, a great deal of time was spent studying an essential substance of nature that was called "phlogiston." It had been observed that when things burned they became lighter in weight, while at the same time they gave off heat and light. The decrease in weight made it seem clear that burning substances lose something of their mass as they give off heat. This released thing that gave off heat was named phlogiston. Another known observation was that if one burned a candle in a closed container, the flame would burn for awhile and then stop. Because the flame stopped with much of the candle left intact, it seemed that depletion of phlogiston was not a likely explanation for why the flame went out. Rather, it was hypothesized that the quality of the air had changed in some way. Indeed, this hypothesis was supported by the observation that if the candle was then exposed to fresh air, it burned again. The conclusion drawn from all of these observations was that for a thing to burn, the air needed to be able to absorb the phlogiston it gave off, and after awhile the burning stopped because no more phlogiston could be absorbed by the air. This is much like dissolving salt in water – the salt will dissolve as it is added (and will, by the way, make the water cooler) up to a point, after which the solution becomes saturated and no more salt can dissolve; however, if fresh water is added, more dissolving can occur.

It was also appreciated that the ability of air to support life had similar properties to its ability to allow things to burn. In other words, if one put a small mammal in an enclosed jar (typically a mouse), it would die after awhile, and the remaining air could not support burning; conversely, if one burned a candle in a container until it burned no more, the air could no longer support the life of a mouse. Amazingly, if one put a plant into air that could not support fire or a living mouse, the plant did quite well, and after a while the air could once again support burning or the life of a mouse. It seemed obvious that both the processes of combustion and of animal life resulted in the release of phlogiston, which is why we breathe (to expel phlogiston from our bodies). Air could only absorb so much phlogiston and thus only support burning or a mouse's life for so long. However, plants removed phlogiston from the air, restoring its capacity to absorb phlogiston and to support mouse life or burning. The phlogiston hypothesis nicely maintained HD coherence; all of the known phenomena could be predicted from the premise that phlogiston existed and that air could absorb only a limited amount.

As scientists began to understand that "air" was actually a mixture of different entities, the study of fractionation of air became a rich area of exploration. When nitrogen was discovered, it was found that neither could flames burn in it nor mice live in it. This was interpreted as creating phlogisticated air; in other words, nitrogen did not have the ability to absorb phlogiston. Some years later Joseph Priestley made the curious observation that the heating of mercuric oxide dephlogisticated air; in other words, air treated by heating mercuric oxide in it was able to absorb more phlogiston, and as such, it increased the burning of a flame and a mouse could live longer in such air than in normal air.[12] The study of phlogiston seemed a triumph, as each new finding about it fit nicely into a coherent theory.

---

[12] Of note, Priestley was also impressed by what he felt when he breathed dephlogisticated air "I fancied that my breast felt peculiarly light and easy for some time afterwards. Who can tell but that in time, this pure air may become a fashionable article in luxury. Hitherto only two mice and myself have had the privilege of breathing it." In 1996 the first "Oxygen Bar" opened in Toronto, Canada, and it has been a trendy thing ever since. Priestley J. 1775. *Experiments and Observations on Different Kinds of Air*. Birmingham, UK: Thomas Pearson.

Ongoing studies of nature demonstrated that some substances, in particular metal substances, actually became heavier when they burned. This was a serious problem for phlogiston theory, as it was difficult to explain why a thing that was losing phlogiston (and giving off heat) would get heavier rather than lighter. A number of arguments were put forth to help explain this, but ultimately (by other studies and famous experiments) it was shown that when things burn they actually combined with something contained in the air. Indeed, by our current understanding, phlogiston didn't exist; rather, the opposite of phlogiston actually existed (which we now call oxygen). A candle didn't stop burning in a closed container because the air could no longer absorb the phlogiston that the candle was giving off; rather, air contained an element vital to combustion (oxygen) and things stopped burning because all the oxygen had been consumed. Heating mercuric oxide didn't dephlogisticate the air; rather, it released oxygen, a gas required for both burning and for life. This represented an inversion of concepts, a paradigm shift no less profound than proposing that the Earth orbits the Sun rather than the opposite.

The point of this story is to focus on the abstract entity that was phlogiston. Like so many things examined by science, phlogiston itself was never directly observed; rather, the effects of phlogiston were observed and served as substantial evidence for its existence. The very notion of phlogiston was basically retroduced (as explained earlier) to guess at the existence of a thing that could provide an explanation for observed effects. As more and more observational evidence became available, it was interpreted in the light of the existence of phlogiston. As has been discussed, the process of retroduction is susceptible to the fallacy of affirming the consequent, i.e., just because if phlogiston existed then it would predict all of the observed effects did not mean that phlogiston must exist (from a logical point of view). However, the idea of phlogiston certainly did exist and worked very well (for a time) to explain the natural world. Many scientists, time and time again, felt they were actually "observing phlogiston" or the absence thereof. Scientists studied the nature of phlogiston. They studied its properties. They could measure its mass as it exited from burning things. They could remove phlogiston from the air by absorbing it with burning mercuric oxide. They could put phlogiston back into the air by burning a candle.

How could scientists study the physical properties of a thing that simply did not exist outside of the abstract human imagination? We do need to pay respect to the broader philosophical notion that there is nothing predictive about our imaginations, no matter how much our posited theories appear to explain; proposing an unobserved entity to explain all manner of observed phenomena does not demonstrate its existence. No scientist has ever directly observed an electron, an atom, or the Higgs boson, yet we state with certainty that they exist, because we can study the predicted effects of their existence. Indeed, we replaced the abstract notion of phlogiston with the equally abstract notion of oxygen, and while the assumption that oxygen exists explains more of our observations than did phlogiston, it is no less abstract. We feel confident studying such things, but at the end of the day each of these things is philosophically no different than phlogiston. Scientists study phenomena, yet they speak in terms of scientific objects. It is essential to remember that many scientific entities that science claims to study will always suffer from the vulnerability of potentially not existing whatsoever. This is also why when one posits initial premises and then builds theories of tremendous predictive power (e.g., Euclid and Newton), one cannot use the success of the theory as evidence that the premises are in fact correct. They may be correct, but they cannot be proven to be so; time and again "known" scientific entities and accepted scientific premises and principles have later been assessed to never have existed at all, other than the idea of them.

In the first two chapters we have described the general logical structure of HD thinking, which is one way in which science has been presented (if not how it is actually practiced). Given the intrinsic flaws in each of its constituent parts (induction, deduction, and retroduction), and the observation that these flaws do little to cancel each other out, it is no surprise that HD itself has all of the aggregate flaws of its own building blocks, if not emergent flaws as well. However, we have only fallen part of the way down the rabbit hole. Humans in general and scientists in particular do not think within the confines of single, individual HD systems. A person's belief constructs are a combined system of perceived causes and effects ranging from the very practical and fundamental (why do I feel hunger, why won't my car start, what are the rules of my world, both physical and societal?) all the way to the ultimate (what are the origins of the universe, what is the meaning of life, why are we here?). We simultaneously hold numerous and complex belief constructs, which influence and affect each other through multiple intersections and cross-effects. In the next chapter we will explore the specifics of systems of thinking that emerge when one combines multiple, smaller HD systems into a larger world view.

3   Holistic Coherence in Thinking, or Describing a System of How Humans Reason and Think

The most exciting phrase to hear in science, the one that heralds the most discoveries, is not 'Eureka!' (I found it!) but 'That's funny...'

– Isaac Asimov

Thus far we have drawn a picture of hypothetico-deductivism (HD), where one can predict outcomes from hypotheses, and where the validity of the hypotheses can be established by investigating whether the predicted outcomes actually occur. If the outcome does occur, then it shows the theory is correct; if the outcome doesn't occur, then it shows the theory is false. This may sound straightforward on the surface, and this is the way science appears to be perceived by many in both the lay public and even by some scientists themselves, but, regrettably, such is not the case. This seemingly straightforward approach differs from how science is actually carried out, and this misperception is both a function of misrepresentation and misunderstanding. The reason for the misunderstanding will be explored later; here it is necessary to define why the testing of hypotheses cannot be as simple as it seems. A nuanced understanding of this issue could not be more essential for a proper understanding of science. It may seem odd, but there are serious problems with determining how evidence confirms a hypothesis and how evidence rejects it – indeed, it is not entirely clear or uncontroversial as to what exactly evidence is or can be claimed to be.

THE ISSUE OF CONFIRMING A HYPOTHESIS

Scientists themselves often refer to data confirming an idea or even proving an idea, but the issue of what constitutes confirming evidence (and how much it confirms) is not clear. The issue of confirmation was most famously dissected by Carl Hempel in his seminal works.[1] Consider the hypothesis that all ravens are black (as Hempel did). From this hypothesis, one can clearly predict that every time someone sees a raven it should have a black appearance. If one sees a black raven, does this "prove" the hypothesis correct? The answer, of course, is no. In order to prove a hypothesis by confirming examples, one would have to observe every raven that has ever existed now, in the past, and in the future. If one were to conclude that all ravens are black after observing every raven except for one, the hypothesis still couldn't be "proven" – that last raven might well be green. This is the same problem of making generalized statements based on finite observations that we encountered in our description of induction in Chapter 1. Thus, the problems of confirmation have similarities to the problems of induction, just with a more obvious practical focus.[2]

However, while observing that a black raven cannot "prove" the hypothesis that all ravens are black, it would seem a very odd notion to state that observing a black raven is no evidence at all in favor of the hypothesis that all ravens are black. But how much evidence should a black raven provide? Carl Hempel introduced a very intriguing notion into the analysis of what constitutes confirming evidence. The hypothesis that "all ravens are black" is depicted in diagram form in Figure 3.1. Because all ravens are black, then the circle defining ravens falls into the set of all black things. Of course, there are many black things that are not ravens (the contents of the light gray circle excluding what falls into the raven circle). There are also a great many things in the universe that are

---

[1] There is a large body of complex philosophical literature on what constitutes evidence, which is outside the scope of this book, but a rich area for the interested reader.

[2] Obviously, the color of ravens is not a terribly practical concern; however, there are many practical applications of this thinking, such as asking if all people with certain symptoms are suffering from the same disease.

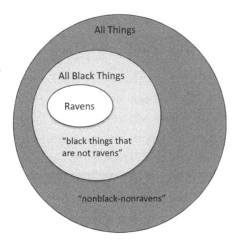

FIGURE 3.1 Graphical representation of the raven paradox

neither ravens nor black (nonblack-nonravens in the dark gray circle not including the "all black things" circle). Hempel pointed out that saying "all ravens are black" gives the very same prediction as saying "all things that are not black are not ravens." In other words, if all ravens are black, and you restricted yourself to only observing nonblack things, then you should never observe a raven (i.e., all things that are not black are not ravens). Because outcomes from both hypotheses are identical, they are seen as equivalent statements.

An apparent paradox arises when one considers that seeing a green apple is making an observation of a thing that is neither a raven nor black. This lends some confirming evidence to the statement "all things that are not black are not ravens" and thus, because they are equivalent statements, also supports the statement that "all ravens are black." If this is correct, then any observation that anyone makes in the world (except for a nonblack raven) is confirmation of the hypothesis that all ravens are black. Thus, any observation whatsoever that does not reject the hypothesis is confirming evidence in its favor. In other words, it's not just a problem that confirming evidence is hard to come by, but also that evidence that confirms is too easy to achieve because all nonrejecting evidence confirms.

Hempel's raven example (often called "the raven paradox") may seem like an abstract and extreme example, but its implications are highly relevant and practical. Consider a detective who is trying to solve the mystery of who murdered a victim on a ship that only had 10 people on board. This is very much like the popular board game *Clue*, in which there are a limited number of suspects. If one can exclude all of the suspects except one, then a confident conclusion can be drawn about that person's guilt. However, none of the evidence that Professor Plum is the guilty party has anything to do with even a single observation about Professor Plum; rather, all of the evidence is regarding what other suspects didn't do. This works because there is a finite and defined set of suspects. Indeed, in the famous words of the brilliant (albeit fictional) Sherlock Holmes, "when you have eliminated the impossible, whatever remains, however improbable, must be the truth."[3] If one limits the possibilities (as in the game of *Clue*), then one can indeed use the raven paradox to prove something. If I know there are only five suspects who could have possibly murdered a victim, and I rule out four of them, then the remaining suspect must be the murderer. However, this only works if I can say with absolute certainty that no one else could have killed the victim, that the victim was actually killed by someone, and that I know for certain that the victim didn't kill herself.

These ideas in no way indicate that a black raven isn't stronger evidence to support the idea that all ravens are black than observing a green apple; it just points out how cheap confirming evidence is and how many things one might be confirming. By Hempel's reasoning, a green apple not only confirms that all ravens are black, it confirms that all ravens are yellow, that all ravens are blue, that all buses are lavender, that all airplanes are orange, and so on. A number of

---

[3] Regrettably, Holmes constantly refers to his thought process as deductive, when in fact it is no such thing. Holmes does not work from axiomatic premises to deduce an outcome that must be the case if his premises are correct. Rather, what Holmes engages in is called inference to the best explanation and is related to retroduction.

solutions have been proposed to the raven paradox as well as detailed philosophical analyses, which are outside the scope of this work. However, there is no consensus on the exact meaning and implications of the raven paradox, or if it even is a paradox at all.

Nevertheless, it seems quite correct to say that if I could observe every nonblack thing in the universe, and none of them are ravens, then any raven that does exist must be black (i.e., all ravens are black). Because the universe contains so many things across a great span of time, seeing a green apple is only an infinitesimal amount of evidence that all ravens are black, but it is indeed evidence nevertheless. How much weight one puts on a minuscule amount of evidence is a separate issue. What Hempel's ideas are really illustrating is that the universe is a conceptually interconnected place, and that one can't really generate a hypothesis about an isolated thing without making statements about all things (albeit of an increasingly tenuous nature the deeper one gets in concept). This is a critical and profound point that needs to be kept in mind in the consideration of how science works in a "holistic" way, as is the focus of this chapter.

The relevance of Hempel's example should not be underestimated, as practicing scientists actually do exactly what Hempel was talking about all of the time, and never more so than right now as advanced computational power and massive quantities of data become increasingly available. The term "big data" has been applied to this field, and it is evolving and defining itself at the current time. But, for example, efforts to find the genetic cause for many diseases are currently underway. In simplistic terms, they consist of sequencing the DNA of people who have a disease and those who don't have a disease, and trying to find the genetic elements that are always present when the disease occurs and never present when the disease is absent.[4] In many cases, the answers are not yes or no, and many genes contribute in complex ways and with nonbinary properties.

---

[4] In many ways, these are the practical application of the methods of John Stuart Mill (Mill's Methods).

That having been said, if one "closes the system" and assumes that the cause for a disease must be found in the DNA sequence of people, if one can rule out every gene except for a single gene, then one can identify that gene as the very likely cause. In other words, it is the equivalent of observing everything in the universe that is nonblack, seeing that none of them is ravens, and concluding therefore that all ravens are black – but without ever seeing a raven. It requires many background assumptions that may not be true, including closed systems of observation, but it is becoming increasingly powerful and real in scientific exploration.

Given that the natural world is likely infinite, or at the very least, far too expansive for humans to observe and experience everything in it, we can never get to any proof through confirming evidence outside closed systems. Nevertheless, assuming closed systems is an essential part of scientific practice. It is necessary to render any idea testable, but this brings with it one of the sources of error in scientific reasoning – that sometimes the closed system assumption turns out to be incorrect. Nevertheless, after accepting this vulnerability, raven paradox–like thinking can render sources of confirmation that are as counterintuitive as they are powerful.

REJECTION OF IDEAS AND HYPOTHESES

If there is any scientific philosopher of whom modern practicing scientists are likely to be aware, it is Karl Popper. In a nutshell, what Popper said is that while one can never prove or completely confirm a hypothesis, one can reject a hypothesis, and it is this practice that helps demarcate science from nonscience. One can't prove that all ravens are black by observing one black raven after another (or one green apple after another) because nature is not a closed system and one can never observe all ravens, all apples, or all things; however, observing a single raven that is not black can soundly reject the hypothesis that all ravens are black. Whereas attempting to confirm a hypothesis always has the problems of induction, rejecting a hypothesis ("falsification" in Popper's words) is deductive in nature, and as such does not suffer the problems of induction. We can reject hypotheses and rule them out. As for theories we hold to be correct, they are never proven but are rather "corroborated" by failed attempts to reject them. The more we fail at trying to reject them, the more corroborated they are.[5]

The ideas of Popper have found their way firmly into the lexicon of professional scientists. Indeed, many academic and professional scientists accept that a hypothesis cannot be proven (although their common use of verification language may not convey such an acceptance).[6] Of those who admit that hypotheses cannot be proven, most will state that they can be rejected and that seeking to reject a hypothesis is the most rigorous way to move forward.[7] Indeed, the idea of ruling out a hypothesis, such that we no longer need to waste time and resources testing it, is a mainstay of how research grants are written, how science is reported, and how academic scientists represent the way that they think. This is one of the hallmarks often attributed to scientific method. A common quote attributed to Albert Einstein is, "No amount of experimentation can ever prove me right; a single experiment can prove me wrong."[8]

Popper used the tendency to attempt to reject hypotheses as a characteristic of science versus pseudoscience and nonscience. Popper

---

[5] Other philosophers have pointed out that while it seems okay to favor a corroborated theory over a rejected theory, it is difficult to justify why a highly corroborated theory is more likely to be true than a less corroborated theory (or an untested theory) without invoking induction; however, this goes outside the scope of the discussion here.

[6] Words like "show," "confirm," and "prove" are tossed around by scientists rather casually, and in my view, recklessly.

[7] Popper used the word "falsification" to describe the rejection of a hypothesis and called hypotheses that were capable of falsification as "falsifiable." This word has a regrettable modern resemblance to the word "falsify," which can indicate that a lie was told and that data was fabricated. This is not how Popper used the word, and in this work the word "reject" or "rejectable" will be used to avoid this ambiguity. "Falsifiable" is used at times when making direct reference to Popper, and in these cases is used as Popper used it.

[8] It is not clear that Einstein said this in these exact words – it is a popular paraphrase probably derived from the actual quote "The scientific theorist is not to be envied. For Nature, or more precisely experiment, is an inexorable and not very friendly judge of his work. It never says 'Yes' to a theory. In the most favorable cases it says 'Maybe,' and in the great majority of cases simply 'No.'"

was particularly impressed with theories from which bold and "risky" predictions followed – that were testable and that would not have been predicted from other theories. If the prediction was not observed, the theory was dead; if it was observed, then corroboration was strong since the prediction would not have been anticipated but for the theory. He used Einstein's theory of relativity and its predictions about light's behavior in large gravitational fields as an example of such a scientific theory.

In contrast, Popper indicated that a theory that could not be rejected by any evidence was a nonscientific theory. A particularly illustrative example that Popper used was theories of psychology that were emerging at his time. Popper considered the theories of both Freud and Adler. As an example, Adler proposed that it was compensating for feelings of inferiority that motivated human behavior. Popper pointed out two opposite examples, the first of a man purposefully drowning a child in an act of murder, the second of a man risking his own life to rescue a drowning child. "According to Adler the first man suffered from feelings of inferiority (producing perhaps the need to prove himself that he dared to commit some crime), and so did the second man (whose need was to prove to himself that he dared to rescue the child). I could not think of any human behavior which could not be interpreted in terms of either theory."[9] Popper's point here is that any human behavior can be explained by the theory, and as such, the theory is incapable of being rejected. If a theory cannot be rejected by anything that can be observed, this doesn't make the theory wrong, it just makes it not susceptible to testing, and as such scientific analysis.

Popper went on to point out the irony that the theories that appear to explain the most (and as such be seemingly the best) were actually the least scientific theories upon which little progress could be made. In his words: "It was precisely this fact – that they always fitted, that they were always confirmed – which in the eyes of their admirers

---

[9] Popper K. 1963. *Conjectures and Refutations: The Growth of Scientific Knowledge.* London: Routledge & Kegan Paul.

constituted the strongest argument in favor of these theories. It began to dawn on me that this apparent strength was indeed their weakness."

This is similar to the debates that sometimes take place between theologians and scientists regarding which approach can better explain experience. The imperfections of scientific understanding, and the inability to explain parts of nature, have been used by those who insist on pitting religious thinking against scientific thinking to claim that science is inferior to religion as an explanation of experience. I would like to acknowledge that, as stated, this view is entirely correct; in general, religion can offer a level of explanation that science can never hope to achieve. If what one is seeking is the ability to predict and control nature, to effect outcomes and modify the human condition, then it is very much the opposite. The difference between explaining nature and being able to predict and control it may seem subtle, but it is profound.

Of course, there are numerous religions with many different properties. In the western and middle eastern worlds, strong adherents to Abrahamic religions point to the fact that doctrine can give complete explanations and answers to the world as we encounter it, or at the very least, there is no experience that cannot be interpreted as an explained result of the belief construct. Even for those things that seem to evade our understanding despite twisting and contorting doctrine, one can always provide the explanation that "humans were not meant to know" that "god works in mysterious ways" that "confusion is god's way of helping us grow and gain faith," and so forth. In contrast, science always leaves us with multiple uncertainties and things for which we have no complete explanation, and in some cases, no explanation at all. In such cases there is no greater goal of nature, there is no virtue or deeper meaning in our ignorance, we are just plain ignorant because we have not yet (and may never) come up with an answer that works. The ability to recognize our ignorance, to embrace it with both arms, and to stare it in the eye is key to scientific thinking.[10]

---

[10] Stuart Firestein explains this in expert detail in his book: Firestein S. 2012. *Ignorance: How It Drives Science*. Oxford: Oxford University Press.

If certainty and explaining the world is your goal, then religion (or a similar approach) is what you should pursue – it seems clear that religion is a much better "explanation" of experience. Such explanatory power is not restricted to deity-driven religions; rather (as explored later) theories of a certain form, even those with the initial appearance of being scientific, can comfortably give explanations for all events. This is the same as Popper's objection to Adler, whose theory could explain why anyone did anything, but could not predict what anyone would do next or in a particular circumstance. In contrast, science does not, and should never claim to provide such level of explanation and certainty. This is not to say that science does not offer explanation, it obviously does, but never in an absolute manner and never in a way that explains everything, at least not as of yet.[11]

The reason many religions can explain everything is because any apparent contradiction (falsification) can be managed by the theology. Note that these theories (or theologies) are explaining everything but predicting nothing. Religion's ability to predict what happens next is pretty poor, despite the immense strength of explaining why something may have happened after the fact. I do not mean to imply that religions and the religious are not inquisitive, critical, or do not at times struggle to understand their experience in the context of their religious beliefs. However, it is permissible in religion to simply offer the explanation that the world is the way it is because a god (or gods) made it so – all is explained at some level. In the words of William of Baskerville: "If I had all the answers, I would be teaching theology in Paris."[12] Light might bend around strong gravitational bodies or not bend, either is explained as god's will. In contrast, Einstein's theory states that light will bend around strong gravitational bodies, and if it does not, then the theory is incorrect. This

---

[11] It is not inconceivable that one day we will "solve the universe" and science will get the whole thing right, but for reasons made clear through the rest of the book, this author is not holding his breath.

[12] This is a line stated by a fictional great scientific mind (also a monk) in Hollywood's adaptation of Umberto Eco's *The Name of the Rose*.

does not make Adler, or certain theologies, or nonfalsifiable theories wrong, but it does render them unable to be a useful theory from the standpoint of prediction and quite incapable of ever being rejected by evidentiary means – thus, from Popper's view, they are nonscientific.

Much analysis has been given to Popper's ideas by other scholars of science. The impact of Popper's ideas is not to be underestimated, but they also face two major hurdles in serving to characterize what science is and how it differs from nonscience. First is the observation that professional scientists, even those who make tremendous progress, often don't act in this way (i.e. they don't focus on trying to reject their hypotheses). The second is a problem that was well known to Popper, and one which he analyzed in his writings – in particular, that hypotheses can never be entirely rejected, even by contradictory data.

WHY HYPOTHESES CAN NEVER BE FIRMLY REJECTED

In my experience, many scientists seem very comfortable with using language and arguments (based on observation) to present the logical rejection of hypotheses whose predictions are not observed in nature. However, baked into the practice of science and its methods are maneuvers that fundamentally address something that is not intuitively obvious and can be a bit striking – in particular, no amount of rejecting evidence can ever entirely rule out a hypothesis. How can this be the case? Is it possible that seeing a white raven cannot be used to reject the hypothesis that all ravens are black?

A number of philosophers have developed a formal understanding of this issue. It is often called the Duhem–Quine problem. Popper himself was clearly aware of this issue, although it is not clear he agreed on the depth of the problem. The problems of rejecting hypotheses were perhaps most broadly articulated by Willard van Orman Quine.[13]

---

[13] Although commonly attributed to Duhem and Quine, Duhem developed this idea in a more particular context and focusing on bundles of concepts, whereas Quine's

Basically, the problem is that isolated hypotheses do not, themselves, make testable predictions. For example, consider a hypothesis that leads to the prediction that all light in a vacuum travels at a constant speed (in this case: 299,792,458 m/s). From this hypothesis, one can deduce the prediction that when one measures the speed of light in a vacuum, one will measure the value 299,792,458 m/s. In order to test this prediction, one would have to engineer an experimental apparatus that had the ability to measure the speed of light. As an example, one would need a source that emitted light, one would need a vacuum of defined length for the light to travel through, and one would need a light detector. Moreover, one would need a timer that could record precisely when the light was emitted and at what time it was detected. So, if one set up such an apparatus and measured a speed of light other than the predicted speed, why couldn't one soundly reject the hypothesis? This seems straightforward enough and logically correct. So, what seems to be the problem here?

If a physicist presented such a rejection, the immediate response by other scientists would be to ask a series of questions, such as: (1) How do you know it was really a vacuum? (2) How do you know your light emitters and detectors were functioning properly? (3) How do you know your timer was working correctly? Each of these is a very fair concern, because the rejection of the hypothesis depends entirely upon the observation being correct, which in turn depends upon each of the above notions.

Quine named such concerns "auxiliary hypotheses," which lurk in the background of all hypothesis testing. In this case, the most obvious auxiliary hypotheses are that the vacuum is a vacuum, that the light source and detector are functioning properly, and that the

treatment of this issue was much more broad, applying to essentially all understanding. To Quine, even "truths" held to be self-evident, such as basic principles of mathematics and logic, where on the chopping block and one might have to solve nature as a whole in order to make specific conclusions about a given hypothesis (Needham P. 2000. "Duhem and Quine." *Dialectica* **54**: 109–32). However, it seems fair to say that both would point out that there is no such thing as a definitive experiment that can compel the rejection of a particular isolated hypothesis.

timer is working correctly. As Quine pointed out, any hypothesis that appears to be rejected by a prediction not coming to pass can be rescued by evoking/rejecting an auxiliary hypothesis. Your observation of light moving at 5,000 m/s does not rule out the hypothesis because it may just be that you didn't measure the speed correctly or what you were measuring wasn't actually light (i.e., one of the auxiliary hypotheses was incorrect). Popper referred to this as reinterpreting the theory "ad hoc" to escape falsification.

Let's return to our previous case of the car not starting when the ignition key was turned. Among the retroduced hypotheses was the idea that the battery was dead. Consider what would have happened if the battery was replaced and the car still didn't start. This could be used to conclude that the original hypothesis was wrong because the battery wasn't the problem. However, for such a conclusion to be drawn with certainty, a number of auxiliary hypotheses (all lurking in the background) each have to hold true as well. A couple of such auxiliary hypotheses might be:

1. The new battery you put in was actually charged and working.
2. You hooked up the battery correctly.
3. It was the correct type of battery.

It should also be noted that auxiliary hypotheses need not be limited to such concrete examples and can be more abstract. In other words, lurking in the background are also additional auxiliary hypotheses – that the battery is not affected by an unusual cluster of sunspots on a given day, by the type of socks the mechanic is wearing, and by whether or not the Chicago Bears won the Super Bowl last year. These auxiliary hypotheses are not something that one needs to create; they already exist in the background of our understanding. Regrettably, like the infinite number of hypotheses we might retroduce, there are also an infinite number of auxiliary hypotheses that one might invoke in order to rescue a favored hypothesis from rejection. In order to isolate and reject a particular hypothesis, one needs to control for an infinite number of different auxiliary hypotheses, which is no more possible than observing every raven that ever has, or ever will, exist – or for that matter every nonblack thing in the universe. So, what about that white raven we saw? Well, how do you know your eyes are really working, or that some filter that changes the color from black to white wasn't put in front of your eyes, or that it was really a raven?

The essential point here is that one can firmly reject a particular "web of belief"[14] as being valid when coherence breaks down – when premises and hypotheses don't support logically valid, reasoned outcomes in the context of background beliefs. However, one cannot identify if the hypothesis, some other background belief, reasoning, or observation is the source of the loss of coherence – and of course, it can be more than one of them. One can regain coherence by modifying any of them. Thus, the addition of observation and experimentation can firmly reject a web of belief to which they are added, but one cannot isolate the hypothesis or the new observation (or any other component per se) as a guaranteed source of the incoherence. Trying to isolate one part of the web of belief is one of the driving sources behind scientific experiments, trials, and studies.

THE DANGERS OF REJECTION AS OPPOSED TO CONFIRMATION

The practice of science is such that previous knowledge claims are always subject to reevaluation moving forward, as more and more information about the natural world becomes available. This is part of the iterative and corrigible nature of science, which corrects itself over time. Thus, accepting confirming evidence (or a failed attempt at rejection) as support for an idea is meaningful, so long as one acknowledges that acceptance is tentative and one continues to seek new information that would be predicted by, or result in the rejection of a theory with ongoing HD testing. Often this does not take the form of performing experiments to further test an idea, but rather coming across information that others have reported, sometimes in a different

---

[14] This term was coined by Quine.

context and for different reasons, but that nonetheless provides information regarding additional predictions of the existing theories – even predictions that did not obviously follow from the theory until the evidence happened to be encountered by someone working on the theory. Although confirmation and rejection may be fraught with problems in logic and practice, so long as one puts reasonable (and not unreasonable) confidence in confirming and rejecting evidence, and one strikes a balance with auxiliary hypotheses, incorrect theories once held to be true are always susceptible to subsequent rejection if ongoing observation fails to line up with what the theory predicts. This, if nothing else, can be found repeatedly in the histories of science. Every theory that has ever been held to be true has been subsequently shown to be false, except for our current theories, which may be on the chopping block tomorrow. As discussed previously, this cannot be done with a logical certainty – making these determinations is part of the practice of science.

Although neither rejection nor confirmation can be certain, rejection of an idea nevertheless seems a much more powerful maneuver than confirmation, as Popper argued. However, the practical flip side of rejection's strength is the dangerous durability of its conclusions. If an idea is incorrectly confirmed, this error will be uncovered by scientists who further study the issue moving forward. In contrast, if something is rejected in a way that scientists accept, then scientists stop studying it. This occurs for a number of reasons. First, most scientists will no longer consider it worth their effort to reevaluate discredited ideas. Second, grant funding agencies are very hesitant to support scientific studies of something that has already been ruled out. Finally, there is a publication bias – if a scientist retests a rejected idea and the rejection holds up (i.e., the idea still appears demonstrably false), it is very hard to get such studies published, as journals do not consider this so-called negative data worth reporting.[15] The

---

[15] Regrettably, publication bias also extends to preventing entirely novel findings that are negative.

publication bias has two equally bad effects: first, it prevents confirming the rejection; and second, it discourages scientists from pursuing studies that reevaluate an existing rejection that may be incorrect.

The danger of rejection is that once an idea is ruled out it is no longer studied and effectively falls off the radar screen. The damage of false rejections is widespread, with profound ripple effects. Because nature is an interlocking system of things that are ultimately of the same universe, the false rejection of an idea can prevent progress in all manner of areas. A false rejection may eventually get revisited if research indicates it is the only way (or best way) to reconcile ongoing conflicts between theory and observation, but returning to a rejected idea is a difficult thing for scientists to do, for the reasons just described.

THE UNDERDETERMINATION OF THEORIES

An appreciation of the interplay of auxiliary hypotheses and an understanding of the raven paradox gives rise to the notion that the natural world consists of interrelated things and ideas. In the words of Quine, we understand the world as a "web of belief," and modifying any one part of the web has widespread effects on other strands and nodes. This notion has been named *holism*. Isolation of one part of the whole separates a part of the world from other interacting parts. This attempt at isolation is exactly why laboratories and simplified models exist – to attempt to limit variables (e.g., control for likely auxiliary hypotheses). However, one ultimately has to consider the system as a whole in order to understand any individual part of it in a natural context, and to appreciate that no matter of isolation can keep the rest of the natural world at bay.

The problem of affirming the consequent prevents us from ever retroducing a single hypothesis with logical certainty; there is always a series of competing hypotheses. However, the problems of holism and auxiliary hypotheses prevent us from isolating any of the retroduced hypotheses and definitively ruling them out. Then how can we ever nail down a single explanation for the parts of nature we can observe?

This concern was explored and developed by Duhem, Quine, and others in the twentieth century and is called "underdetermination."[16] Basically, underdetermination is a formal statement of the problem stated earlier, that no amount of data will ever be sufficient to reduce all possible explanations to a single hypothesis, rejecting all others. Thus, theory is always "underdetermined by data." Moreover, it's not just a problem of being unable to reject all of the retroducible hypotheses except the right one; it is also that one cannot completely rule out even a single hypothesis, because no matter how damning the rejecting evidence may be, any hypothesis can survive the rejecting evidence by changing a background belief or altering an auxiliary hypothesis (as pointed out by Duhem, Quine, Popper and others).

Because science typically restrains itself pretty severely by the existing web of belief, underdetermination is much less of a problem in the day-to-day practice of science. The number of testable hypotheses simultaneously in play in a particular field are usually quite limited. Typically, science does not suffer from a problem of numerous hypotheses, all of which equally explain all the data. This is not to say there will not be competing hypotheses in science – there always are. However, when this occurs, it is often the case that not even a single hypothesis can be generated to explain all the existing data, and competing hypotheses succeed and fail to explain the data in different ways.

Why is this the case? The existence of the web of belief (e.g., our existing base of understanding built upon centuries of compiled evidence) constrains what hypotheses can even be considered, as any hypothesis that violates the web of belief is not consistent with all the known data. Likewise, the number of auxiliary hypotheses that do not violate some part of the web of belief are limited. Errors in the web of belief can be tremendous barriers to progress, because scientists are

---

[16] Stanford Encyclopedia of Philosophy. 2009 (updated 2017). "Underdetermination of Scientific Theory." http://plato.stanford.edu/entries/scientific-underdetermination/#FirLooDuhQuiProUnd

hesitant to start modifying any part of the web unless they are compelled to do so by new evidence or by a theoretical impasse that they can't circumvent without questioning parts of the web previously held to be true. In this way, science is fairly conservative, both discouraging leaps of innovation but also preventing reckless quackery. However, keep underdetermination in mind as you read on.[17] When science goes wrong (and it does so, tragically at times), it is often the theoretical concern of underdetermination rearing its very real and ugly head. A misunderstanding of auxiliary hypotheses leads to all manner of error. As the saying goes, "assumption is the mother of all screw-ups." An assumption is just another way of saying that auxiliary hypotheses are present that are unjustified and potentially incorrect.

TESTING OF AUXILIARY HYPOTHESES IS NORMAL HUMAN THINKING

Scientists mostly employ typical human thinking (induction, deduction, and retroduction). Testing auxiliary hypotheses, the issue brought up by Quine, is likewise normal human thinking.

Consider a scenario where your doctor diagnoses you with cancer. It's a shock, and you struggle to adjust to the new information. But what if the diagnostic test for the cancer gave the wrong answer? What if a clerical mistake was made, and your name was put on someone else's specimen? You have just challenged the auxiliary hypotheses (or background assumptions that make up your web of belief) required for your cancer diagnosis to be correct (e.g., the specimen really came from you, the test was run properly, the pathologist read the results correctly, the pathologist's report was labeled with the correct patient's name, etc.). Again, while we do not normally refer to such ideas as auxiliary hypotheses, it is clear that there are countless background assumptions that must be correct in order for a primary idea to be true, and it is normal human thinking to question

---

[17] Please read on.

them when scrutinizing the main idea. When faced with a conclusion of which we are skeptical, we challenge it by asking: But how do you know (insert background assumption here)? In contrast to our tendency to question auxiliary hypotheses, humans also have an amazing capacity to ignore them when the main hypothesis is exciting or desirable.

Prior to the allied invasion of Sicily in the Second World War, a body dressed in the uniform of a British Royal marine washed up on a beach in Spain. His body carried the identification of Major Bill Martin. In his pocket were papers documenting that the invasion would occur in Greece and Sardinia, not Sicily. This information was passed up to the German high command and caused them to prepare for the defense of Greece and Sardinia, leaving Sicily vulnerable to invasion. However, as with all webs of belief, multiple auxiliary hypotheses were hidden within it. Some of these auxiliary hypotheses were that the body was really a major in the British Marines, that the orders were authentic, that the orders were still valid, that the allies had not changed their strategy since the orders were issued, etc. In fact, the body was that of a British man who had died from accidental ingestion of rat poison in England, and, in order to mislead the Germans, his body was put into a British military uniform with false papers stuffed in his pocket and deposited in the sea off the coast of Spain. This subterfuge by the British was named "Operation Mincemeat." Some members of the German command were skeptical and questioned the auxiliary hypothesis that Bill Martin was really an authentic major who had been killed in action. Luckily for the allies, the British had created a false identity for Major Martin and arranged circumstances that ultimately led to German acceptance of the orders as authentic. The German error was to accept the auxiliary hypothesis that the body was really a British military major.

Auxiliary hypotheses are hidden in all of our thinking, and challenging auxiliary hypotheses is a common way to scrutinize our conclusions. In the practice of science, the maneuver carried out to address auxiliary hypotheses is called using "controls." Let's say a scientist is performing an experiment to test the hypothesis that a particular virus is responsible for a certain disease. A test is run on a patient with the disease and no virus is detected, leading to a rejection of the hypothesis that this virus is responsible. For such an experimental outcome to be accepted as valid, the scientist must typically include a "positive control," which consists of adding a known quantity of the virus to an otherwise negative sample. This positive control essentially assesses the auxiliary hypothesis that the test is working properly and is able to detect the virus. In the event that the virus was detected in every sample analyzed, or at least at an unusually high rate, then one would need a negative control (a sample that was known to not contain the virus) in order to assess the auxiliary hypothesis that the test doesn't mistakenly detect the virus when it is absent. Thus, when one reads scientific literature, the logical underpinnings of the requirement for controls is the ever-present issue of observations being dependent on the validity of a number of auxiliary hypotheses.

While many scientists may hold strongly to the idea that hypotheses can indeed be firmly rejected, they nevertheless spend a great deal of time and energy running controls precisely because rejection cannot free itself from being mistaken due to a false background assumption (e.g., an auxiliary hypothesis not being correct). Thus, the practice of science takes into account the problems of rejection, but scientists may be unaware of the logical reasons for doing so, and the impossibility of actually rejecting a hypothesis is seldom articulated in the scientific literature or even acknowledged by practicing scientists.

Although not often referred to in formal terms, the problem with rejection is certainly understood in everyday thought. When debating an issue, the common idiom (all other things being equal) is, essentially, stipulating that none of the infinite background beliefs and auxiliary hypotheses are different. Thus, isolating a single idea by holding all other things equal is essentially a thought experiment that addresses the problem of auxiliary hypothesis and background beliefs/assumptions.

Like the concepts of auxiliary hypotheses themselves, the use of what scientists call controls is a common human practice. If you replaced your car's current battery with a new one and your car still didn't start, it would be normal to test the new battery with a voltmeter to see if it is sufficiently charged or putting it in a different car to see if that car starts with it, which is "controlling" for the auxiliary hypothesis that the new battery is functional. Thus, the normal process of human troubleshooting includes the testing of other background assumptions (or auxiliary hypotheses) through the use of maneuvers that control for such issues.

Despite the inability to reject hypotheses due to an infinite number of auxiliary hypotheses needing to hold, this does not mean that the virtues of rejection are to be ignored. The more auxiliary hypotheses one accounts for (i.e., controls) in the light of evidence inconsistent with what is predicted by a hypothesis, the more likely a hypothesis is to be incorrect. Moreover, because rejection is similar to deduction in nature (whereas confirmation is inductive), rejecting hypotheses carries greater weight than does confirmatory evidence. In other words, our observations can give us more information when an idea is wrong than when an idea is correct. Thus, an emphasis on rejecting evidence over confirming evidence is appropriate, so long as the issue of background assumptions is not ignored, and rejecting evidence is not seen as a way to certainty in the real world, with all its messiness and infinite auxiliary hypotheses.

CHANGING BACKGROUND BELIEFS ALTERS COMMONSENSE CONCLUSIONS

Common sense is a much lauded property of clear thinking people. However, common sense is simply the reasoned predictions of a coherent web of belief grounded in a body of common background assumptions found among a population of people.

In 2015, an outbreak of Zika virus infections in Brazil was linked with an increased rate of babies born with microcephaly. If you had been a pregnant female admitted to a hospital in Chicago at this time, and you needed a blood transfusion, common sense would dictate that you should insist on blood that had been screened for Zika virus and had tested negative. This seems obvious and would not require much consideration. However, had you insisted on blood that had tested negative for Zika at that time, you would actually have maximized your chance of Zika infection compared to untested blood. Why would this be the case? The reason is that early on, when testing capacity was very low, the decision was made to only apply the limited testing resources to blood collected from regions where Zika was known to be endemic (e.g., Puerto Rico).[18] As such, the only way to get a unit of blood that tested negative for Zika was to import blood from Puerto Rico, where Zika infection was actively occurring. The test for Zika is good, but like most tests, it does not detect 100% of infected units. The chance of the Zika test not picking up Zika in an infected blood product from donors in an endemic region was higher than Zika being present in an untested unit of blood collected from other regions.[19] As such, insisting upon blood that tested Zika negative would result in getting a unit from regions where Zika was present and increased chances of infection. If you didn't have the background information about the nonuniformity of testing, which most people would not have had, then insisting upon blood that tested negative was the correct choice. However, this approach goes from the best choice to the worst choice simply by changing one background assumption.[20]

---

[18] This example is taken from issues regarding the blood supply in the United States, and as such, Puerto Rico was one of the only regions where blood was collected and where Zika was endemic.

[19] The probability of someone in the Midwest having an active Zika infection was not zero, as they could have become infected through travel or from intimate contact with an infected person. However, travel is an exclusion criteria for donation, assuming the donor remembers and answers the screening question correctly, and infection between people is uncommon. So, overall, the probability is vanishingly small, but not zero.

[20] Later on, as testing resources became more available, all blood in the United States was tested for Zika. Whether this is a wise use of resources remains a matter of debate.

I recently encountered a more commonplace example in the seemingly simple task of trying to determine what my daughter prefers to eat. While preparing for a visit to my folks in Chicago, my parents asked what items my daughter liked to eat at that time, so they could have some in the house. I gave them a list of the things that she had asked that I put in her lunch each day. This is just good common sense; my daughter asks for a certain yogurt drink to be in her lunch every day, and so it is a food item she prefers. Each day I would put the drink in her lunch, and each day she would bring her lunchbox home empty – evidence suggesting that the drink had been consumed and then discarded. However, my daughter later revealed to me that she really detested this particular yogurt drink.

Of course I asked her why she insisted on having it in her lunch if she didn't like it. The answer was that while she didn't like it, a friend of hers at school loved it, and my daughter would trade the yogurt drink for a kind of muffin her friend always had in her lunch. Of note, her friend didn't much like the muffins but loved the yogurt drink. I asked my daughter, "why don't we just get the muffins ourselves, so I can put them in your lunch?" The answer was that trading food items at lunch was fun and part of the societal experience of being a kid at her school. Initially, my web of belief had simply not included a background understanding that there was a vibrant barter economy going on at my daughter's school at lunchtime. Absent this information, it seemed reasonable to conclude that my daughter actually enjoyed consuming the food item she insisted on having each day. Here the data did not change with an altered background belief – either way my daughter was still asking for the yogurt drink each day; however, the meaning of that data and the conclusions I could draw from it was entirely different.

RESCUING HD COHERENCE BY ATTACKING PREMISES, OBSERVATIONS, AND AUXILIARY HYPOTHESES

Importantly, the effect of auxiliary hypotheses is not restricted to assessing the main hypothesis. One may also change his or her interpretation of evidence or conclusions about a situation, based on a change in different background information. A change in any part of the web of belief can result in a series of compensatory alterations in other parts.When I was a medical student, I spent 4 weeks training in the psychiatric inpatient ward at the hospital attached to my medical school. I was paged to evaluate a patient who had been sent to the psychiatric service. After speaking to her for a while, it became clear to me that she was greatly distressed by her belief that someone had implanted a clock in her chest cavity. She knew it was there because she could constantly hear the "tick-tock" of the clock. I marked down on my notes that she was suffering auditory hallucinations, took down her history, and then performed a physical exam. As part of the exam, I noticed a sternotomy scar on her chest (she had had some kind of chest surgery). When I placed my stethoscope on her chest, to my surprise, I heard a loud and consistent tick-tock in her chest just as she had described. After further questioning, she revealed that she had undergone open-heart surgery some years previously (a fact I had failed to elicit and she had failed to volunteer when I took her history). I was eventually able to hunt down the details, and it turned out that she had an artificial "ball and cage" valve implanted in her heart to fix a valve defect. My initial retroduction that her perception of a tick-tock in her chest was caused by an auditory hallucination included the background auxiliary hypothesis (or assumption) that there was no actual tick-tock in her chest, an assumption that was wrong. This patient was misinterpreting the sounds, and suffered from a lack of insight as to the cause, but my thinking had contained an assumption that had turned out not to be the case. When I heard the sound myself, I had to modify my belief construct to maintain HD coherence, so that my premises once again predicted what I had observed – she actually did have a tick-tock in her chest, just not from a clock.[21]

[21] Of course, I could also have rescued coherence by concluding that we both had auditory hallucinations or by many other modifications of auxiliary hypotheses.

After shopping in a store, you push your cart of groceries to your car, approaching it with HD coherence. In other words, you have a web of belief regarding your car and expectations regarding what you will experience. You push the button on your key to open the trunk but it doesn't open. This is in contrast to your prediction that, given the premise that this is your car and the key opens it, pressing your key should cause the trunk to pop open. Your coherence has now been disrupted; the trunk that should have opened has not. By classical ivory tower HD, you would use the data (which is clear, sacred, and cannot be altered) to reject one of your hypotheses. Thus, you challenge the hypothesis that this is really your car. If it is your car, you can predict that your license plate will be on it; you look, and indeed a different tag number is there. In light of this new evidence, you now reject the hypothesis that this is your car and coherence is restored (you would not expect your key to open someone else's car). The problem has been solved and you go in search of your actual vehicle. Although this example is clean and logical, there are often many other ways that such an example can play out.

Let's explore three ways by which you could easily restore coherence regarding your car. First, you could challenge your observation itself (e.g., the trunk actually opened, but you didn't notice). Alternatively, if you push the key button over and over again, and it opens on the third try, then coherence has been restored and the data were just incorrect (or at least not reproducible). However, if the trunk still doesn't open despite many attempts, you might try to attack a different point in your coherence; you could challenge one of your background assumptions (auxiliary hypotheses). For example, your assumption that your key is working – after all, the battery in your key may be dead or the battery in your car might be dead and is not receiving the key transmission. Indeed, coherence would be regained if your key was in fact malfunctioning, because what you have observed would once again follow from your premises and background beliefs. Even if the key is no longer transmitting a signal, you can still predict that it should manually open the lock. You manually insert your key into the keyhole and attempt to turn the key to open the trunk; however, the key doesn't turn, so coherence remains disrupted. You now turn your attention to your background hypothesis, that this is your car. To test this, you look at the license plate on the car, which is a different number than you remember your plate being, leading you to conclude that you have mistaken a similar looking car for your own. Any of these maneuvers might have worked in restoring coherence (e.g., the trunk might really have opened, the key might have lost its radio signal but still worked manually, or this might not be your car).

The importance of this particular example cannot be exaggerated. The simplistic common narrative of HD thinking that is often applied to science is that predictions can be deduced from hypotheses, data can be collected to test the predictions, and if the data do not support the prediction then one can soundly reject the hypothesis. As stated before, this is a normal part of human thinking. How often do you hear the phrase, "Well, we know X isn't true because of Y." However, the reality of the situation is simply that neither everyday thinking nor scientific thinking really works in this way. Rather, as the previous examples illustrate, logical coherence can be maintained in at least three major ways: (1) by modifying the theories, (2) by questioning the data and observations, or (3) by changing background assumptions (modifying auxiliary hypotheses). One tries to establish and maintain the greatest level of agreement between theories and the observations that can be predicted from the theories, changing different parts of the equation. This is far from the portrait of science that is often painted, in which clear and unambiguous data allow the logical and methodological rejection of theories; rather, all three of these maneuvers can be carried out (sometimes simultaneously) to maintain HD coherence.[22]

---

[22] A fourth maneuver that can be taken to try and restore HD coherence is to challenge one's reasoning (e.g., deductive-type logic) that leads to the predictions from the premises. One can reassess reasoning and change the predictions that follow from hypotheses and data without changing anything else. Scientists certainly think

As a classic historical example of this in the "hard sciences," let's look at Sir Isaac Newton's theories of gravity and planetary motion (Newtonian mechanics). Newtonian mechanics has been one of the greatest and most successful theories in the history of science; indeed, it has been heralded by many as a quintessential example of scientific triumph and intellectual achievement. By making some base assumptions (which he called laws), Newton was able to deduce a mathematical system that described how gravitational forces resulted in planetary motion in our solar system and the entire universe – quite an accomplishment indeed![23] However, despite its success in predicting almost all the relevant data collected by scientists at the time, discrepancies were subsequently found. First and most famously, the orbit of Uranus was found to deviate from that which is predicted by Newtonian mechanics. Thus, HD coherence was lost, as an observation did not line up with what was predicted by the theory.

Astronomers were in agreement that the motion of Uranus was not consistent with Newton's theory.[24] The ability of various scientists to check and recheck the same natural phenomenon over time remains a strength of science in this regard; thus, rejecting the data was not an acceptable way to maintain coherence – the observation wouldn't go away.

At this point, a strict application of rejection based on disconfirming evidence should result in Newton's theory simply being rejected as false – no matter how much confirming evidence there is, one failed deductive prediction is sufficient to reject. Indeed, one *could* simply have rejected Newton's theory, but the web of belief was

---

reasonably, but outside of mathematical disciplines, they seldom think in formal deductive constructs.

[23] Arguably, Newton retroduced his laws, because he knew much about planetary motion when he came up with them and then developed the math from them that lead to known outcomes.

[24] To those of you who are now giggling at your sophomoric pronunciation of Uranus, your behavior is consistent with that of most scientists (and most other people, for that matter).

strong – the theory had been so successful in so many ways, that there was no rush to throw it out. Rather, an attempt was made to restore coherence while protecting both the theory and the data by challenging a background auxiliary hypothesis, in particular, that there were no hitherto undiscovered planets. It was therefore speculated that there was a large unknown body in space that was pulling Uranus out of its predicted path due to a strong gravitational force.

Indeed, based on calculations and predictions put forth by Urbain Le Verrier using Newton's equations, Johann Gottfried Galle discovered Neptune on September 23–24, 1846.[25] The discovery of this previously unknown planet that was affecting the path of Uranus was an additional triumph for Newtonian theory, but it also serves as another illustration that data contrary to that predicted by a hypothesis does not necessitate the rejection of the hypothesis. In this case, one of the infinite background hypotheses (i.e., that there are no additional undiscovered planets) was challenged by the auxiliary hypothesis of Neptune's existence. Since the auxiliary hypothesis led to its own testable prediction, which was assessed experimentally (Neptune could be seen with telescopes), the wheels of HD thinking moved forward and HD coherence was restored. This latter point is critical, as Popper pointed out that introducing an ad hoc assumption to rescue a hypothesis that was untestable (e.g., that didn't make its own predictions) renders the whole hypothesis unrejectable – in this case, they were able to look for the new planet that was predicted.

The great triumph of Neptune's discovery was seen as a validation of both Newtonian mechanics and science itself. Le Verrier had predicted the location of Neptune to within one degree, an amazing accomplishment. It is thus no surprise that Le Verrier approached another disagreement between Newtonian theory and observed data with similar enthusiasm. It had been observed that the perihelion

---

[25] It has also been claimed that John Quincy Adams had made the same calculations.

precession[26] of Mercury deviated from what Newton's system predicted; the deviation was small but consistent, and multiple astronomers found the same result in their observations. Again, HD coherence was disrupted. The findings were not deducible from Newton's premises and theory. Coherence couldn't be easily restored by rejecting the observation, since one could check and double-check the observation and it wouldn't go away. Coherence couldn't easily be restored by rejecting the theory, because so much evidence supported Newtonian mechanics. As with the discovery of Neptune, coherence was restored by challenging a background assumption (again, the assumption that there were no additional undiscovered planets). Le Verrier calculated that a hitherto undiscovered planet between Mercury and the Sun (which he called Vulcan) would lead to the prediction of Mercury's measured perihelion precession. In 1859 Edmond Lescarbault observed Vulcan and communicated his observation to Le Verrier, who announced the discovery with great enthusiasm in 1860. Thus, once again, HD coherence was reestablished by adding the auxiliary hypothesis of an additional planet.

The astute reader may, at this point, be concerned at never having learned much in school about the mysterious first planet of our solar system, Vulcan (other than in Star Trek, but that Vulcan is not in our solar system). The reason for this is that no one other than Lescarbault could consistently observe such a planet.[27] Thus, reestablishing HD coherence by discovering Vulcan was less firm than when it had been done by discovering Neptune. Indeed, to this day, there has been no confirmation that Vulcan has ever existed, and despite the fact that its existence would have restored coherence to Newtonian mechanics, data simply don't support this conclusion. Thus, HD coherence was once again lost.

---

[26] Planets rotate around the sun in an elliptical orbit, and the elliptical paths themselves rotate slowly over time, which is called a perihelion precession. see https://en.wikipedia.org/wiki/Apsidal_precession for illustration.

[27] Multiple "sightings" of Vulcan have been reported over the years, most recently in 1970; however, none of these claims has held up to observational scrutiny by astronomers.

So, what is a thinker to do? If we accept that Mercury's perihelion (as observed) it is not predicted by Newtonian mechanics, and if no auxiliary hypothesis that withstands scrutiny is forthcoming to reconcile Mercury's perihelion (e.g., the Vulcan claim), it could just be that no one has been creative enough to think of the right auxiliary hypothesis (perhaps our method for measuring perihelions is flawed). Alternatively, the weak point may be the underlying theory itself. However, Newton's theory had been so successful and predicted so much with stunning accuracy that it was hardly an easy target – in the web of belief, there were multiple and very strong connections to empirical evidence that were anchoring the theory in place.

Coherence was ultimately recovered in 1915 with Albert Einstein's development of the theory of relativity, which is not accounted for by Newtonian theory. Relativity theory predicts Mercury's perihelion precisely. Thus, shockingly, in this latter case, the theory (Newtonian mechanics) itself was incorrect in spite of its great success and had to be modified in light of the data. Ultimately, it seems that Newton was wrong, but the balance of HD coherence was regained.

Thus, in two separate cases regarding Newtonian mechanics, HD coherence broke down and in two separate cases it was restored by modifying different parts of the web of belief. In the first case, the auxiliary hypothesis that there were no undiscovered planets was modified. In the second case, the theory itself was rejected. While these seem like very different acts, they are identical when viewed through the lens of modifying some part of the web of belief to seek HD coherence. But this raises a critical question, if one can modify any part of the web of belief and maintain coherence, what are the rules for when to modify what?

HOW TO CHOOSE WAYS OF MAINTAINING COHERENCE

The fact that coherence can be maintained in multiple ways is a major problem – how do we know which is the right way? Let's return to the example of your key not opening your car trunk. Your key didn't work, either electronically or by inserting it into the lock, so you move your attention to a base premise and challenge the hypothesis that this is your car. You notice the license plate is not yours and so you conclude this is not your car, which restores coherence. However, you could equally restore coherence by maintaining that this is your car but that someone switched the license plate while you were shopping, reprogrammed the key's signal, and rekeyed the lock. If you look into the car and don't see your belongings but rather the presence of unfamiliar items, all of your experiences could still be explained by changing the single hypothesis (that this is your car). Alternatively, you could rescue the hypothesis that this is indeed your car if you add the additional assumption that someone opened your car and replaced your belongings with those of someone else while you were shopping, changed the license plate, rekeyed the lock, and changed the signal from your key. Logical coherence is achieved equally by a single change in your hypothesis (this is not your car) or by the four listed changes required to retain belief that it is your car.

The comedian Steven Wright made light of this very problem when he said, "The other day somebody stole everything in my apartment and replaced it with an exact replica." Walking into your apartment and seeing everything exactly as you remember it is equally coherent with nothing having changed or a space alien replacing all of your stuff with identical copies. One would typically favor the first explanation because the second explanation is so unlikely. However, the important point here is that neither explanation is logically superior, as both maintain HD coherence. In both normal human thinking and scientific practice, one often uses the principle of Occam's Razor to help sift through this problem (e.g., the simplest explanation is the most likely[28]). In other words, there is no

---

[28] This quote is attributed to the medieval scholar, William of Occam (also Ockham). There is actually no record of him ever having said this, although it is somewhat consistent with his sentiments that the hypothesis necessitating the fewest assumptions should be favored. This idea was expressed as early as Aristotle, but was made most famous by Occam: "*Numquam ponenda est pluralitas sine*

need to make things more complicated than required to maintain coherence; however, there is no particular reason why things need be simple, even though we would like them to be. That being said, because each of our individual beliefs is linked into a web of belief, and the web carries the weight of much empirical evidence at endless intersections, there is some rationality to maintaining coherence by the means that causes the fewest unnecessary changes to the web.

COHERENCE AS A FILTER TO THE WORLD

Maintaining a logical coherence between our base beliefs, our background assumptions, and our observations is one way that we navigate the world. We pass through most of our day not taking note of the majority of things we observe. One may drive to and from work, pass thousands of cars, and have little sense of their characteristics. However, if one were to see a single car floating in the air above the highway your attention would immediately be drawn to it, because this is inconsistent with your experience and opposed to your belief constructs and background assumptions. This is why magic tricks are so appealing – they violate our sense of how things should happen. When we see a person levitated on stage, our first response is to assume we're being tricked in some way. We evoke auxiliary hypotheses, e.g., that the floating body is hanging by some wire or is supported by some pedestal we cannot see. This is precisely why the magician passes a hula hoop over the body of the floating subject, so as to reject such auxiliary hypotheses. If we can be deprived of all our auxiliary hypotheses by the magician, we are left with the explanation that the magician really can levitate a person. Most of us remain of the opinion that we're being tricked, but just can't figure out how.

---

necessitate" or "Plurality is never to be posited without necessity." More simple and requiring fewer assumptions are not exactly the same thing, and the distinction can be very important.

Evoking auxiliary hypotheses and background beliefs to maintain coherence fuels the discord in our current political systems. Using American politics as an example, everyone may acknowledge that Robert Mueller is carrying out an investigation of Donald Trump and his associates. Those on the far right may evoke the auxiliary hypothesis that there is a leftist conspiracy (or even a deep state) and this is all a "witch hunt." Those on the far left may evoke the auxiliary hypothesis that Donald Trump really has broken all manner of laws and that the only way the investigation can fail is if the Trump establishment intervenes. Still others in the "Q movement" may evoke even more complicated background assumptions around a more convoluted landscape. Each of these requires different auxiliary hypotheses, which may be more or less likely, but each view is entirely coherent with its own web of belief – and every web of belief connects to the fact that Mueller is carrying out his investigation.

There are all manner of conspiracy theories and complex belief structures held by a great many people. While many beliefs may seem extreme, they are by no means incoherent. Indeed, a truly exceptional conspiracy would leave no trace of itself whatsoever, and thus the lack of any evidence of a conspiracy is the best evidence of a truly sophisticated conspiracy. While we may reject such ideas as being unlikely, in many cases our rejection cannot fault the logic of conspiracy advocates, as their beliefs are just as internally coherent as any other belief. Even delusional people can maintain coherence with excellent efficiency. (Some forms of psychosis may have a total break with reality and a disorganization of thinking; however, a delusional person can be somewhat different, i.e., just because someone believes that a demon or the CIA is controlling their thoughts, it doesn't mean they are not thinking logically – albeit with premises that may not seem likely or well grounded.)

A main point of this chapter is that maintaining HD coherence is a common vehicle used by laypeople and professional scientists alike. It is a fiction, in both the sciences and everyday life, that one gains new data and then rejects or accepts hypotheses in a clean and clear fashion. The everyday example of the car not opening, the scientific example of Newtonian mechanics, and even fringe conspiracy beliefs can be equally maintained by altering any part of the web of belief, from hypothesis, to background assumptions, to questioning the observation itself. There is no way to logically pin down any one part of the equation, unless you assign certainty to the other parts of the equation, and there is no basis for doing that. Understanding how one maintains HD coherence is an essential building block in generating a working definition of science and defining how it can be distinguished from nonscientific activities. It will be argued that maintaining HD coherence is a necessary component of science; however, as illustrated previously, HD method alone cannot be sufficient to distinguish science from other approaches, as it is commonly found in all manner of thinking. The main point is that while an accepted observation cannot reject a hypothesis, observation can compel a change in some part of the web of belief (of which the hypothesis is one part), and in doing so it can at least make a change to our thinking (if not some kind of progress). This requires that what appears to be HD coherence, really is, as will be explored in later chapters.

THE NEED TO DEFLATE SCIENCE TO A REALISTIC STATUS BASED ON COHERENCE AND THE WEB OF BELIEF

It is essential to deflate science to a state more realistic than the common hyperbole of what science is and how confident we are in its claims. Because scientific claims depend upon HD coherence, which is built upon a system that uses induction, deduction, and retroduction, any knowledge claim made by science suffers from all of the shortcomings entailed in each of these tools (as detailed in the first three chapters), as well as the problems of holistic beliefs and underdetermination defined in this chapter. While it will be argued in subsequent sections that the nature of scientific claims is indeed different from nonscientific processes, science and the facts that it puts forth have all of the previously discussed problems of normal human reasoning. In my view, the scientific method should be viewed as an iterative system that has made great progress in recent centuries and will hopefully continue to do so. However, there is no scientific proof or definitive statement that is not subject to the above fallacies.[29]

When politicians, think tanks, and lobbyists justify nonaction about global warming or environmental regulation because science has not "definitively shown" a problem exists, or because there is no "scientific proof" that a particular problem exists, or because science remains "uncertain" about a particular problem, this causes immense damage, because it is based on a fantasy of what science could be in an alternative universe where logic worked differently and absolute truth was possible. In such a universe, lack of absolute proof might be a good reason not to believe something. But we don't live in that kind of universe, or at least we don't experience the universe that way as humans, and therefore great damage is done through our misinterpretation of how scientific knowledge claims work (e.g., allowing people to disavow environmental and health crises because there's "no proof" that they exist).

In some cases, the issue of "scientific uncertainty" is a cynical maneuver by groups who purposefully exaggerate "doubt" to justify damaging products and activities[30]; in other cases it likely reflects genuine confusion as to what science really is. In either event, so long as we live in this universe and as humans, science must be viewed through a more realistic lens – as an intrinsically flawed process that will not achieve certainty of anything. This in no way implies that there isn't something special about science that makes its

---

[29] Proofs may be achieved in abstract mathematics and pure theory untied to natural observation; however, once one engages with the real world all the above shortcomings apply.

[30] Oreskes N, Conway EM. 2010. *Merchants of Doubt*. London: Bloomsbury Press.

assessments of the natural world different from other approaches, but the person who waits for scientific certainty will wait forever.

Science always has a base level of incoherence, of internal disagreement of theory and observations. Scientists use increased coherence as a guide for modifying different parts of the web of belief when it might be justified; however, scientific thinking tolerates a great deal of incoherence. In the previous sections, the theoretical problems with retroduction, affirming the consequent, and the underdetermination of theories shows us that there are an infinite number of hypotheses to explain any observation or observations. As such, the problem seems to be how to narrow them down to a testable number. However, this is seldom the problem that practicing scientists face, at least in the context of broader theories. In contrast, our best scientific theories seldom (if ever) predict all of our observations, and if they do, then just wait until we have a few more observations.

For most broad bodies of observational data, there is no single theory that predicts them all, and as such, far too few coherent hypotheses rather than too many. This is not necessarily because no comprehensive theories exist, but likely because at any given time, a certain amount of observation is flawed, parts of the web may remain undefined or undiscovered, and the reasoning linking parts of the web may be misguided. Thus, one cannot pin down other parts of the web and focus on theory alone. Any part of the web may be flawed, and not knowing which is which prevents progress of a certain nature, but this does not equate to preventing progress altogether. Rather, it is iterative and imperfect, but progress nevertheless.

Reconciling that incoherence may motivate much scientific thinking; however, scientists are fairly comfortable with a base level of incoherence. It is for this reason that you will typically only hear nonscientists reject the theory of evolution because it can't explain every last specific detail of the fossil record or the lack of some predicted fossils. It is for this reason that you will typically only hear nonscientists reject global warming theory because it doesn't predict weather patterns with certainty or because some anomalies exist in the data. Essentially, all theories have anomalies and things they can't explain. Only nonscientific theories explain everything perfectly and without exception. As was pointed out by Popper, their apparent strength is actually their weakness – they are so flexible as to explain everything but then so malleable as to be essentially useless for any other purpose than abstract explanation, with no ability to predict.

However, having laid this groundwork, we must keep the following question in mind: If science gives less certainty than other systems of belief, is there any reason to prefer scientific claims? We are also justified in asking: If nonscientists are more stringent than scientists in rejecting theories as untrue when they fail to predict nature, why would we choose to listen to scientists over nonscientists? Both of these are very important questions that we must address. But before we can answer these questions, we must continue to develop our definition of science so that we understand what it is and what it is not. Having defined the reasoning tools and coherent basis that scientific and nonscientific thinking have in common, we will turn our attention in the next section to particular ways in which science differs from and can be distinguished from other modalities of thinking.

PART II

4   How Scientific Reasoning Differs from Other Reasoning

To abdicate from the rule of reason, and substitute for it an authentication of belief by the intentness and degree of conviction with which we hold it, can be perilous and destructive...

– Peter Medawar

In the early 1950s, a group called the Seekers formed in a suburb of Chicago, based on the belief that they were receiving messages from a greater intelligence through a process called "automatic writing." Automatic writing occurred when a medium (in this case, a woman named Dorothy Martin) entered a trance-like state that allowed her to write out channeled messages from a greater being called Sananda. Martin's hand would basically take on a mind of its own and messages from Sananda would come forth on paper. An entire belief construct was derived from these messages, including an understanding that they were coming from a faraway planet named Clarion and that UFOs from Clarion were frequently visiting Earth.

At one point in early 1954, Martin received a message that the Earth was going to be destroyed on December 21, 1954, by a cataclysmic flood emanating from the Great Lakes. This prediction galvanized her group of followers into action; they quit jobs, sold property, and upended their lives in an effort to prepare for (and survive) the flood. One of Martin's followers, Dr. Charles Laughead, was a medical doctor, who had previously been a physician at Michigan State University. He began to publicize Martin's messages and lent some academic credence to her ideas. Unbeknownst to Martin or any of her followers, a group of psychologists and sociologists who were interested in doomsday cults had infiltrated the group for the purposes of making a sociological study. This resulted in the excellent book,

When Prophesy Fails, by Leon Festinger, Henry Riecken, and Stanley Schachter. It provides detailed and invaluable insight into the inner workings of this particular belief system and its adherents.[1]

It seems likely that most individuals would not categorize the Seekers as a group of scientists, and it does not appear that they were engaged in scientific activities. However, in the context of our evolving definition of science and hypothetico-deductive (HD) coherence as a manner of thinking, it is not abundantly clear at first glance that what the Seekers had been doing was not science. Sananda, although an alien that the group believed had previously been on Earth as Jesus Christ, was not assigned a divine status, but rather was simply treated as a greater intelligence from an advanced civilization that had greater technologies than humans do; thus, communications from Sananda can be considered natural phenomena.[2] Specific data were collected from Sananda in the form of automatic writing, and the corpus of received messages was compiled into a growing body of "lessons." Martin's automatic writing was not isolated data; indeed, fellow channelers generated similar information (albeit by channeling different sources), and it was noted that the group specifically put more weight on messages that had been "corroborated"[3] from multiple sources, and hence there was some form of consistency (if not reproducibility) to the phenomenon.

Perhaps most importantly, just as Karl Popper had required of science, this group's hypotheses led to risky and specific predictions, and testing these hypotheses was not only achievable but essentially unavoidable. In addition to the apocalyptic prediction for December

---

[1] In the book, to protect personal privacy, Dorothy Martin was called "Marian Keech" and her location was described as being in Michigan.

[2] At several points during the Seekers' activities, additional channelers came forth who wove the messages from Sananda into a Christian narrative and thus did imbue their thinking with a divine element, but these ideas were short-lived and did not appear to heavily influence the core belief structure of the Seekers.

[3] Somewhat ironically, this is the same word used by Karl Popper when he presented his definition of science.

21, 1954, numerous other predictions were made regarding members being visited by aliens and UFOs rendezvousing with the group to pick up members at particular places and times. The Seekers showed up at the specified rendezvous for UFO pickup, but no UFOs arrived. December 21, 1954, came and went without any incident or great flood. Some unwitting human visitors who came by the house the group occupied were suspected to be extraterrestrials; they were questioned extensively but ultimately were determined not to be aliens. As more and more predictions failed to occur, some members of the group began to change the basis of the predictions and the particulars of the belief construct to explain the disconfirming evidence. For example, after December 21 failed to be apocalyptic, some decided that they had misinterpreted the messages or that the date for Earth's destruction had actually been moved. Others postulated that all of their preparations had staved off the apocalypse in some unspecified way. It was also proposed that the world had actually ended but they could not perceive it because they were in a better place. Eventually, the failed predictions (disconfirmations) became too numerous for the belief construct to handle and the group disbanded, with most ultimately rejecting the belief construct. Martin never stopped believing in Sananda or Clarion. She moved first to the Peruvian Andes, later to Mount Shasta in California, and then to Sedona, Arizona, where she died in 1992. She continued automatic writing for the rest of her life and founded the Association of Sananda and Samat Kumara, using the name Sister Thedra.

At first analysis, it is not entirely clear why the activities of the Seekers are any different from what scientists typically do. Based on observable data that could be shared with colleagues (the messages from automatic writing), the existence of Sananda was retroduced. Sananda was not directly observable, but the effects of Sananda's existence were easy to see in the form of channeled writings. The belief construct made specific predictions that were testable. When the predictions did not come true, some challenged how the data had been interpreted (the world ended, but we could not see it). Others evoked auxiliary hypotheses to rescue the main hypothesis (the world was saved by the adherents' activities). Others modified the belief construct to incorporate the disconfirmations. Enough disconfirmations ultimately arose to cause the rejection of the belief construct, except by a few dedicated adherents. The Seekers' process seems to match our working description of the scientific process quite well.

We can then ask: Why do the Seekers not fit into the category of science? Is what the Seekers were doing so very different from how astronomers evaluated Newton's theories of planetary motion, when orbits and perihelions did not behave as predicted? Were not the Seekers retroducing hypotheses (just as Peirce and Whewell had described), making risky and testable predictions (just as Popper had insisted on), testing the predictions through observation (as any empiricist would require), and modifying the web of belief at all three levels (challenging data, modifying auxiliary hypotheses, and modifying premises), just as Quine would describe holistic science? Quite often in science, certain adherents of a theory (often those who came up with it) never reject it despite overwhelming evidence against it. This, too, was found to be the case with the Seekers, as neither Martin nor Laughead ever gave up their beliefs.

WHAT LOOKS LIKE HYPOTHETICO-DEDUCTIVISM OFTEN IS NOT

Many have questioned HD as a defining characteristic of science, in part because groups like the Seekers seem to be acting like scientists and engaging in HD thinking, but in my view, this is a misinterpretation of their activities.[4] To see this clearly, a juxtaposition to HD in action in scientific progress is very useful. As an example, one can analyze the genesis of theories about infectious disease. One pattern that has been repeatedly observed by physicians throughout the ages is that disease outbreaks often cluster around areas with poor living

---

[4] Others have questioned HD thinking as a model for scientific practice as many professional (and highly accomplished) scientists appear not to practice it. This is explored in later chapters.

conditions and public hygiene, such as areas with decomposing animal matter and standing pools of water. Prior to the seventeenth century, "miasma theory" was a predominant explanation for disease outbreaks. The basis of miasma theory was that illness was transmitted by "foul vapors," or poisons in the air. The foul vapors could be released from decaying animal remains and thus could emanate from standing water (which usually had muck in it), decomposing animals, and/or refuse. The power of miasma theory is that it gave a plausible explanation for why people tended to get sick in clusters in close proximity to certain water sources, dead and decaying bodies, and where basic sanitation was lacking. In other words, the theory predicted exactly what was observed in the world. The miasma theory had been retroduced to be consistent with observed nature (i.e., patterns of disease). If the miasma theory was correct, then one would predict the clustering of disease around standing water and decaying animal matter, which is exactly what was observed.

So, what was the problem with miasma theory? Well, one could derive predictions of disease from miasma theory in addition to those clustering around water and decomposing animal matter. For example, since the source of disease is foul vapors from decomposing material, then according to miasma theory, disease could not be transmitted from one living being to another, as living animals do not decompose. However, this prediction was shown to be incorrect. In the early 1800s, the silk industry was failing due to a plague among silkworms. Agostino Bassi demonstrated that he could transmit the disease from one silkworm to another through inoculation (injecting fluid from a sick worm into a healthy one). Subsequent experiments by numerous scientists, most famously Louis Pasteur, demonstrated that germs (small microscopic entities) from sick animals could be grown and then injected into healthy animals, resulting in the same illness that had afflicted the original sick animal. Thus, germ theory (first proposed in some form at least as early as the 1500s) emerged as the predominant theory of illness, because it explained the observed world better than miasma theory. Of course, it could have been that both germs and miasma lead to disease, but it was subsequently shown that rotting organic material did not spontaneously give rise to life forces, an outcome predicted by miasma theory and upon which miasma theory depended.[5]

How, then, is the example of miasma theory any different from the Seekers' belief construct regarding Sananda, Clarion, and UFOs? In both the case of miasma theory and the Seekers, a cause that was not itself directly observable was posited to be responsible for an observable effect; i.e., an ill-defined and unobserved thing called "miasma" caused observable disease, while a better defined but also unobserved thing called "Sananda" caused observable automatic writing by Dorothy Martin's hand. There was corroboration in both cases: Many other physicians documented disease clusters around foul water, and other members of the Seekers channeled similar predictions. Both miasma theory and the Seekers created testable predictions, and both were ultimately refuted and abandoned because the predictions did not hold up to observation of the natural world.

However, a fundamental difference between the two is the "deductive" part of the word "hypothetico-deductive." In the case of miasma theory, the following outcome was induced from observations of nature.

Induction: Sick individuals cluster around sources of rotting organic matter (this includes stagnant sources of water).

From this observation, the hypothesis of miasma was retroduced. Miasma was a disease-causing entity that was generated from rotting organic matter, which then made those exposed to it ill.

---

[5] It should be noted that many scientists of the time had reported the observation of spontaneous generation; however, it was later shown to be due to contamination with new living things. Thus, this is the case where observations themselves were reinterpreted over time by further challenging auxiliary hypotheses (i.e., the rotting matter isn't contaminated with new living things). This is Quinian holism in action, the maintenance of coherence by modifying different parts of the web of belief, both theory and observation.

Consistent with being a valid retroduction, the retroduced hypothesis deductively leads to observed nature.

Premise (hypothesis): Rotting dead matter is a source of miasma.
Premise: People exposed to miasma have higher rates of disease.
Deduced outcome: Therefore, people exposed to rotting dead matter will have higher rates of disease.[6]

If one stipulates in their hypothesis that miasma is the only cause of disease, then one can also deduce that a person cannot become ill merely from exposure to a sick living animal because living things do not give off miasma (i.e., in the absence of rotting miasma-generating material).

In the HD method, general principles of nature are induced from observations, and hypotheses are retroduced, from which one can deduce the observed outcomes (and optimally additional unobserved outcomes that one can subsequently test).[7] The role that deduction plays in this process is in the prediction of outcomes that must be the case (or cannot be the case) if the hypotheses (premises) are correct and auxiliary hypotheses hold (all other things being equal). Given the miasma theory of disease, one can deduce that the transmission of illness from one living animal to another cannot occur, as living things do not give off miasma. If such transmission does occur (observation is correct) and the deduction was correct, then the source of error was in the premises themselves or in assumed background information (auxiliary hypotheses). If HD coherence is lost, then regaining it requires challenging either the hypothesis itself, one of the

---

[6] Note that we are not saying that all people exposed to miasma will get sick – clearly not 100% of people living near rotting material become ill; rather, it just increased the chances of getting sick. A number of philosophers, most notably Hempel, recognized that HD constructs had to be able to handle probabilistic predictions, and as such, could only be tested by observing populations, which is the case here.

[7] A number of philosophers of science have made the ability to deduce as-of-yet untested predictions a requirement for a valid scientific retroduction and a requirement for science, because without it, hypotheses cannot be tested further or distinguished from each other.

background auxiliary hypotheses, the validity of the deduction, or the correctness of the observations – in other words, one is compelled to change the web of belief to regain HD coherence.

The reason that Dorothy Martin's writing and the hypothesis of Sananda cannot conform to the requirements of HD coherence lies in the deductive component. Certainly, it was a verifiable observation that words came forth from Dorothy Martin's pen. Many people observed the pen writing and the printed "lessons" were evidence of that writing having occurred. From this evidence, Dorothy Martin retroduced the existence of Sananda, which was a hypothesized cause that could explain the observed effect. At this point, we must acknowledge that the Seekers' hypothesis consists of two different parts. First is the hypothesis that channeling of Sananda was responsible for the writing, and second was that the writing (and thus Sananda) had some ability to foretell certain future events.

However, while these two hypotheses are certainly tied together for the Seekers, they need not be. One can separate the fact that the writing occurred from the question of its predictive power. For example, the writing could simply be coming out of Dorothy Martin (the Sananda Hypothesis is untrue), but still have predictive power because Dorothy Martin has clairvoyant abilities or is herself a space alien. Alternatively, the writing really could be coming from Sananda, but Sananda may have no ability to predict future events at all. Or perhaps Sananda knew the future but was purposefully giving false predictions for motivations that were beneficent, maleficent, or indifferent. Of course, both parts of the hypothesis could be false, and, like any retroduction, there are an infinite number of alternate hypotheses that can be evoked to explain the data (e.g., the writing was coming from a Soviet mind-control ray that was pretending to be Sananda, and the predictions were a trick to induce a panic in American society). The fact that the predictions did not come true can firmly reject the second part of the Seeker's hypothesis (all other things being equal).

The above nuances notwithstanding, it is essential to consider that the Seekers' were treating the hypothesis as a single entity – as such, a failure of the predicted outcomes of the lessons could firmly reject the whole construct, that the lessons came from Sananda, who was making accurate predictions.[8] The reason that this construct does not have a HD nature is because an intelligence imbued with free will is the cause of the observed effects and predictions.[9] Why is this a problem? From the standpoint of miasma theory, given the premises that miasma comes from decaying animal matter, and that exposure to miasma causes disease, then one can deduce that people who live around decaying animal matter are more likely to get sick (all other things being equal). If, people living around decaying animal matter do not get sick at a higher rate, there may be a number of explanations available through the evoking of auxiliary hypotheses; however, one of them *is not* "miasma changed its mind and decided not to make people sick in that particular case."

Even if Sananda was a real intelligence on a faraway planet with the ability to transmit through Dorothy Martin's pen, the content of the messages cannot be deduced. Sananda may make one prediction, have a change of mind, make another prediction, or make no prediction at all, and all of these are consistent with the presence of Sananda. There is also no deduction that necessitates the messages will come forth from the channeler's pen, as Sananda may simply choose to stop communicating. The whole basis for HD is that hypotheses can be challenged if their predictions don't hold, precisely because the predictions must occur if the hypothesis is true and auxiliary hypotheses are kept constant. However, in the case of Sananda, one need not even change auxiliary hypotheses to explain a different outcome. All other things being equal, Sananda may decide

---

[8] There is a logical complexity of language at play here. A number of philosophers have explored the complexities of language of this type, perhaps most notably Bertrand Russell, and the interested reader is encouraged to explore this area.

[9] What free will is, how it works, who has it, and if it even exists is a debate of great complexity with a large literature available to the interested reader. It is clear that humans experience free will – the ability to choose – but some hold that this is a misperception.

to do this or do that, with no alterations in the rest of the web of belief. Sananda is an intelligence with free will. Since Sananda can choose what to do then no predictions need occur, even if Sananda is real. Therefore, the failure of any prediction to come to pass cannot be used to reject the idea of Sananda. It is for this reason that the Seekers were not engaged in HD thinking, even though they might appear to have been, even though specific testable predictions were coming forth from their group. This is not to say the Seekers were stupid or that they were not reasoning well – it is just that the structural basis of their system does not have a deductive component. They maintain systemic coherence but not HD coherence. Without a deductive component, one cannot compel a change in the web of belief through new observation. If one is unable, under any circumstances, to compel such a change, and furthermore, if one is unable to reject a hypothesis (even when holding all auxiliary hypotheses fixed and stipulating correct observation), then one cannot carry out science.

Of course, ultimately, the inability to reject hypotheses is not what plagued the Seekers. Most of them attempted to rescue the hypothesis of Sananda through evoking of auxiliary hypotheses, but ultimately, they did indeed reject the hypothesis when too much disconfirming evidence accumulated. Only the truly fanatical held fast to the Sananda hypothesis despite all the disconfirming evidence. However, what the Seekers were rejecting is that the predictions did not hold up – they became frustrated at the lack of pragmatic utility and thus abandoned the belief.[10] However, unlike the abandonment of the belief in the existence of miasma (or phlogiston for that matter, see Chapter 2), the Seekers could not reject the existence of Sananda – who, unlike miasma or phlogiston, may simply have had a

---

[10] If one held a very strict, pragmatic view of science, one might argue that scientific theories are useful or not based solely on whether they work; the Seekers were doing this. They had a theory, they wanted to see if it worked, it didn't, so they rejected it. However, even strict pragmatist models of science require HD constructs to make consistent predictions when all other things are equal; the Seekers were lacking this.

change of mind – *all other things being equal.* The Seekers simply abandoned that the practice of listening to the predictions was useful.

WHY STUDY OF SUPERNATURAL COGNITIONS CANNOT BE A SCIENCE: THE REQUIREMENT FOR DEDUCIBLE OUTCOMES

As a principle of science, explanations of the natural world can only involve natural causes and natural outcomes. In other words, explaining that the Earth is getting warmer because it is God's will or due to Satan's effects, or due to the emanations of the ever-divine Flying Spaghetti Monster as per the Pastafarian religious order, cannot be a scientific statement. Once divine powers are evoked, the discussion is outside the realm of science. But why must this be the case?

On the surface it appears that scientists, theologians, and spiritualists all do the same basic thing. They observe what occurs in nature, they claim the existence of causes to explain such effects, and then they go out and experience more of the natural world under the belief constructs they have embraced. In this process, a theologian postulating that a god exists seems no different than a physicist postulating that dark matter exists; both are retroducing a cause for an observed effect. Neither the god nor dark matter are directly observable by the human senses, and neither can be measured directly even with our best instruments. However, the effects of such entities existing can be readily found in human observation. Should dark matter exist, it would explain much of the observable behavior of heavenly bodies by astronomers. Should a god or gods exist, it would explain a great deal of the workings of the world and even the existence of the world itself. Moreover, I am unaware of anyone ever having been spoken to by dark matter.[11] In contrast, numerous people have experienced being directly spoken to by a god or gods, and

---

[11] Given the complexities of human psychology, I believe that it is very likely that somewhere in the world someone perceives that they are indeed being spoken to by dark matter.

countless people alive today have perceived a sense of supernatural beings. Whereas scientists need specialized equipment to measure many of the phenomena they study, a theologian or spiritualist needs only their own senses to pray, meditate, or feel the energy of the universe and the divine. One could even argue, and some have, that religious and spiritual thinkers base their beliefs more on their direct experience of the world than do scientists, and thus have more evidence of the existence of gods and supernatural entities than do scientists for some of their favored hypotheses. So, what is the difference between science and religion? How can science be capable of supporting the existence of dark matter but not capable of supporting the existence of a god? If science is a system that evaluates the world based on observation and evidence, it seems there is much more direct, experience-based evidence of a god than of dark matter. This is true for most scientific beliefs – billions of humans have likely directly felt the presence of some spiritual entity, but only a relative handful of humans have directly observed most scientific phenomena.

One reason that science rules out studies of supernatural cognitions is because of the requirement for deduction as a component of this process. If dark matter actually existed, and holding the correct background beliefs and auxiliary hypotheses constant, then one could deduce, with essentially logical certainty, that particular physical properties in the universe would be observed. Sufficient subsequent investigation of the physical makeup of nature that failed to find the predicted properties could essentially rule out the existence of dark matter (or at least force some change in the web of belief – if not ruling out dark matter, then compelling an alteration in an auxiliary hypothesis). However, the existence of a god or spiritual universal energies has no specific deductive consequences that could lead to particular predictions. It could be argued that the very existence of the world is a deducible consequence of the presence of a god; but couldn't the god have simply decided not to create the world?

Let us consider the premises of typical Western monotheistic systems; that God made the universe, and that God is loving, benevolent, all-knowing, and all-powerful.[12] It is often argued that the observed world does not follow from such a premise. Bad things happen to many people who appear to be good people and who follow all of God's rules. Natural disasters (which an all-powerful God could control) destroy cities and kill thousands of people. Plagues and pestilence afflict many humans. Indeed, if the statement that a loving, all-powerful, benevolent, and all-knowing god would not allow bad things to happen was deducible from the premise, then the presence of such calamities would allow for the rejection of the existence of such a god. But a god or gods can be capricious, can be angered, and can behave in unpredictable ways. A god may be testing us, even for our own good, in a way that we cannot appreciate or comprehend. So, aren't these explanations simply evoking auxiliary hypotheses – like any good scientist would do? In fact, they are in a way, but such auxiliary hypotheses have no deducible consequences themselves – so the web of belief remains unlinked to observable outcomes, at least in any deductive way. Moreover, one can have two entirely different outcomes to the universe, even with all other things being equal, by a capricious god simply changing his or her mind, with no necessary cause for that mind change – assuming of course that the god him- or herself has free will. If there is no cause to an altered effect, then there is no HD.

Why is it okay to rescue Newtonian mechanics from its failure to predict celestial motion by positing dark matter (that has no additional deducible consequences that we can currently observe), and it is not okay to rescue the theory of God by explaining horrible things happening to good people as "God works in mysterious ways" or as "God gave humans free will, and as such, some humans will do evil to other good humans." What about Quine's holistic objection that even in the hardest of sciences, hypotheses can never be rejected, as one can always rescue the hypothesis by changing an auxiliary hypothesis (e.g., the web of belief)? Doesn't Quine's point regarding the inability

---

[12] These terms are often used by Western monotheistic systems and are put forward only as one example of how God is defined.

to reject even scientific hypotheses make the inability to reject theory of God the very same thing as science? This apparent problem disappears if we change the criteria from a requirement to be able to reject a hypothesis to the less ambitious requirement of being able to force some change to the web of belief, of which the hypothesis is a part.

Whereas positing dark matter does not lead to deducible consequences that we can currently observe (other than the motions of celestial bodies that dark matter was retroduced to explain), there are consequences that are testable if our technology becomes advanced enough. It is not inconceivable that we could someday send a probe to an area of space hypothesized to contain dark matter. This scenario is consistent with the history of science. Indeed, none of the additional predictions of Einstein's theory of relativity were testable at the time he formulated the theory; rather, testing it would depend upon waiting for a solar eclipse to occur and later the invention of new and novel technologies over the next century capable of assessing the deductive consequences of the theory as it was stated. However, moving back to the issue of dark matter, even without new technologies, the unfolding of the natural world may test dark matter on its own and be susceptible to our passive observation. For example, if observations of celestial body motion in the future came out a certain way, it would violate deducible consequences of dark matter as it is currently conceived, and force a change to the web of belief. However, no outcome of the natural world can ever rule out that God works in mysterious ways. No outcome of the natural world can ever rule out that God is testing us. While it is reasonable to conceive that we may eventually develop technology to directly probe dark matter, it is not reasonably conceivable that we will create a technology that allows us to test the mind of God. No occurrence whatsoever can compel a change in the web of belief of God. As such, while both science and religion appear similar in how they explain the world in the context of their premises, they differ at least in this very fundamental way. If no outcome of experience can, under any circumstance, compel a change in the web of belief, then no science can be done.

Remember the example of your car not starting and how the HD method was used to troubleshoot the problem? If one suggested that the problem lay in the battery, or the starter motor, or the ignition, then one could test each of these directly. However, if one suggested that the car didn't start because the spirit of Elvis was preventing the car from starting, how could one test such a hypothesis? There is no observable result that one could deduce from such a premise, and therefore, there is no way to assess the idea. Of course, one could pray to the spirit of Elvis and then see if the car started – this would be a kind of test of the theory. However, as Popper pointed out, if the car didn't start it would be because the spirit of Elvis didn't want it to start, and if it did start it would be because Elvis had been placated by the prayer, and his spirit now wanted the car to start. Any outcome maintains the existence of Elvis's spirit controlling the car, and no outcome can compel a change to the web of belief.[13] If all outcomes support the idea, and no evidence can disconfirm it, then one is not testing a hypothesis.[14]

These ideas do not speak to whether gods, demons, or Elvis's spirit exist or have effects on the world; they may or may not. Rather, this simply speaks to the idea that science cannot assess such claims. It is for this reason that supernatural cognitions are not allowed in the scientific realm, not because they are disliked as a concept but because there is simply nothing to be done with them scientifically. On this issue science is not a skeptic; rather, science must simply be silent. If claims of scientific evidence are being made by those studying supernatural cognitions, then science may reject that the claims have "scientific merit," but this is different than claiming they have "no merit" – it just isn't scientific. Conversely, those who would deny the existence of God, even those who are professional scientists, are not doing so from a scientific standpoint. At most, they may be

---

[13] To my knowledge, Popper never spoke directly on the issue of Elvis, per se.

[14] This discussion does not explain why science can't study supernatural things that are not based on a cognition. This is explored more deeply later.

objecting to the use of nonscientific thinking to make certain claims, which is a different issue. However, if they are denying the existence of God, they are not doing so scientifically; this is not an issue that science can address.

It is of greatest importance that we recognize that this position of science was not chosen; rather, it was forced upon science by the very nature of how science works and the role HD coherence plays in shaping the web of belief. The inability of science to study cognitions (supernatural gods or nondivine beings like Sananda) is due to the ability of the cognitions to capriciously change without any alteration in any other part of the web of belief. Another way of saying this is that the cognitions are not rule governed. Under the same initial conditions, and with the same hypotheses and auxiliary hypotheses, the cognitions can do different things.[15] The broader requirements of

---

[15] This argument might seem to support a conclusion that human psychology cannot be a science, as humans are generally conceived of as a cognition that has free will. First, we should note that the study of human psychology focusses on the question of how the human mind functions, not trying to figure out if humans actually exist based upon effects that might be attributable to an otherwise invisible human cognition. However, stipulating that humans do exist and do have free will, the above argument would still suggest that psychology cannot be a science since it deals with cognitions. This likely would be true if the study of humans was restricted to analysis of a single human individual out in the natural world and with no experimental control, much like a single supernatural cognition or the single case of Sananda. However, the science of psychology does not typically build or test its theories on single individuals; rather, it juxtaposes different experimental groups of people under conditions in which variables can be isolated, and with careful attention to probability distributions. For example, human psychology science may conclude that a certain percentage of humans will suffer post-traumatic stress disorder (PTSD) after being exposed to a certain type of trauma, but predicting a priori exactly which humans will and won't develop PTSD may not be possible. Likewise, when subjected to tests of reasoning and cognition, humans (as a group) have predictable tendencies; however, one can only tell what humans tend to do, and not what a single specific human will do in a given circumstance. Thus, there can be scientific study of the rules of human cognition as they apply to a population, but not precise prediction of one person's behavior. This issue of how probability affects the ability of scientific theories to predict and control is explored later in this text. It is worth considering that we seldom (if ever) have a population of gods to study. There are polytheistic systems, but the gods are typically each different. However, if we had access to several thousand examples of a god of a certain type, or even several thousand examples of different members of the Sananda species, and

being governed by rules, at least to some extent, in order to be the object of scientific study is discussed in broader detail later in Chapter 5.

MISTAKEN REASONING BY HUMANS (AND SINCE SCIENTISTS ARE HUMANS...)

Formal deduction has very specific requirements and properties, as described in Chapter 1. While a HD model of science has some descriptive relevance to science, in many cases, the link that practicing scientists make between hypotheses and prediction is certainly a type of reasoning, but cannot conform to the formal definitions of deduction. This is especially the case in more complex systems – the more complicated the system, the more difficult it is to maintain any real deductive coherence. For a highly controlled and specific experiment, one may have a very distinct HD construct, but once one brings the variables of the real world to play on the system, introducing numerous complexities, the likelihood of maintaining formal deduction becomes less and less. Nevertheless, the systems must maintain the ability to use distinct reasoning to make testable predictions, and if the predictions do not hold, it must force some change in the web of belief by the reasoning that is being used. One might call such a model hypothetico-predictive instead of hypothetico-deductive, to reflect that reasoning leads to predictions, even if not formal deductions. We shall continue to use the term HD, while recognizing that formal deduction may not always be at play. However, reasoning is certainly at play. This means that mistaken human reasoning would be one way in which the web of belief might be flawed, and by which HD

we could undertake a repeat study of how they act under different controlled circumstances, and recording probabilities of outcomes and results, then it may be possible to bring supernatural or alien cognitions into the realm of scientific study. Even repeated access to a single god, over and over again, to carry out a study of how the god responded over time might make some progress. Of course, there would still be the requirement that the underlying behavior of the being was at least somehow rule governed and that the beings would cooperate with our studies. Nevertheless, if these requirements were fulfilled, it is possible that science could commence.

coherence may be lost. A prediction that follows from a hypothesis may have failed to be observed not because the hypothesis was incorrect or observation is flawed, but because our reasoning was incorrect in concluding that the prediction actually followed from the hypothesis. But we don't need to worry too much about this, because humans (and especially trained scientists) are highly skilled thinkers with good powers of reason and the ability to think clearly, right?

Imagine you are lucky enough to be chosen for a new game show called "Win All the Prizes" and you are the sole contestant on the stage. You are shown three semi trucks in a parking lot, one of which is packed full of awesome prizes worth $1 million; the other two trucks have no prizes in them. You have no idea which truck has the prizes. You are given the chance to choose any of the trucks, and you choose truck number one. The host of the show (who knows which truck contains the prizes) then opens up truck number three to show you that it's empty. One of the two remaining trucks contains the prizes. You are now given the option to stick with your initial guess (truck one) or to switch your guess to truck two. What gives you the best chance of winning the prizes? Take a minute and think this over carefully!

If you have come to the conclusion that it really doesn't matter what you do because there is a 50/50 chance of the prizes being in either truck, you have come to the same conclusion as most people. In a study performed in 1995, 87% of people (out of 228 subjects) chose to stick with their original choice.[16] However, you (and the 87% who agree with you) would in fact be completely incorrect if you did so. The reality of the situation is that you will win the prizes two-thirds of the time when you switch to the new truck and only one-third of the time when you stick to the first truck you chose. This is a version of a famous probability problem called the "Monty Hall problem."[17]

---

[16] Granberg D, Brown, TA. 1995. "The Monty Hall Dilemma." *Personality and Social Psychology Bulletin* **21**(7): 711–29. doi:10.1177/0146167295217006

[17] There is an excellent description and history presented at: Wikipedia. n.d. "Monty Hall problem, Sources of confusion." https://en.wikipedia.org/wiki/Monty_Hall_problem#Sources_of_confusion

Don't worry if you're in the 87% group – you're in good company. Indeed, after the publication of the Monty Hall problem in *Parade* magazine in 1990, close to 10,000 letters were received by the magazine, some in quite derisive language (including about 1,000 from readers with PhDs), arguing that the column was incorrect. Reflecting on this puzzle, psychologist Massimo Piattelli-Palmarini stated "even Nobel physicists systematically give the wrong answer, and...are ready to berate in print those who propose the right answer."[18] Many famous mathematicians have refused to accept the answer, until shown mathematical simulations illustrating the effect. As a humbling side note, pigeons that are repeatedly exposed to the Monty Hall problem learn to always switch choices.[19] Indeed, sometimes trial-and-error based learning defeats the power of our analytic human brains.

A detailed explanation of why switching choices results in winning two-thirds of the time is beyond the scope of this work and can be found elsewhere.[20] However, a simple explanation is that by choosing one out of three trucks at random, you have a one in three chance of winning, which is what occurs if you pick (and stick with) your first choice. Remember, the host of the game show knows which truck contains the prizes. If you guessed correctly in your first pick, then the host can open either of the remaining trucks, but if you did not guess correctly then the host will never open the remaining truck with the prizes, always choosing the one that is empty. The idea that the two remaining trucks (after the host opens an empty one) have a 50/50 chance presupposes that the trucks are equivalent, but such is

---

[18] vos Savant M. 1996. *Power of Logical Thinking.* New York: St. Martin's Griffin, p. 15.

[19] Herbranson WT, Schroeder J. 2010. "Are Birds Smarter Than Mathematicians? Pigeons (*Columba livia*) Perform Optimally on a Version of the Monty Hall Dilemma." *Journal of Comparative Psychology* **124**(1): 1–13. doi:10.1037/a0017703. PMC 3086893. PMID 20175592

[20] Wikipedia. n.d. "Monty Hall problem, Solutions using conditional probability and other solutions." https://en.wikipedia.org/wiki/Monty_Hall_problem#Solutions_using_conditional_probability_and_other_solutions

not the case. After the empty truck is opened, the remaining truck (which you didn't pick and wasn't opened) has survived a selection process, and thus you have more information on that truck than on the one you initially picked. Because the door you picked is never opened, you have no additional information on that truck. By switching trucks, you essentially have the opportunity to look in two trucks instead of one, and will win two-thirds of the time. In other words, if you pick a truck and stick with it, you only get to guess once. However, by picking one truck and then switching, you get to guess twice and look in two trucks.

The relevance of the Monty Hall example to our discussion of scientific thinking relates to how HD coherence is maintained. In earlier chapters, we emphasized that HD coherence can be kept by modifying hypotheses, modifying interpretation of observation, or by modifying background assumptions (auxiliary hypotheses). However, we failed to include in this list that one can also modify reasoning. That is to say, a hidden requirement for HD coherence is that the reasoning is valid. However, coherence can be destroyed if one uses flawed reasoning. Regrettably, as with the Monty Hall problem, humans don't always reason correctly. One can deductively demonstrate with mathematical proofs that one should switch trucks; the less formal but common reasoning that it doesn't matter only has the appearance of being logical and correct – but to many people that appearance is very strong.

Sadly, commonsense reasoning can often lead to deductively invalid arguments. Part of scientific training is to learn to recognize valid logical arguments. However, few scientific curricula contain any formal training in logic. Even with such training, it is a humbling observation (as illustrated by the Monty Hall problem) that even well-trained scientists and mathematicians can fall into the trap of poor reasoning. However, as science is an iterative and self-corrective process, this is remedied over time and in this case by experimentation (e.g., many of the advanced mathematicians who would not accept the Monty Hall problem changed their opinions when they saw the simulations – basically an experimental result).

THE ROLE OF HEURISTICS IN MISTAKENLY VIOLATING DEDUCTIVISM

Navigating the world is an incredibly complicated task. One way that we appear to have addressed this problem is by the formation of certain cognitive "rules of thumb" that we apply to problems or scenarios of particular types. The rules of thumb that the human brain tends to use have been termed "heuristics" and are best associated with studies carried out by Amos Tversky and Daniel Kahneman, who were awarded the Nobel prize for their work in this area. A heuristic is reflexive thinking that human cognition uses instead of approaching a problem analytically; it is rule of thumb thinking. Cognitive psychologists agree heuristics occur. However, why they occur, in what real-world settings they manifest, and the implications of their existence remains a matter of debate.

Kahneman and Shane Frederick subsequently described a process by which heuristics affect reasoning without our awareness of the heuristic being there, a process called "attribute substitution." Basically, when presented with a complex problem, the human mind substitutes a simpler problem for which an answer is easily obtainable and may do so without the person being consciously aware of it. The field of heuristics – of understanding common processes of human cognition – is a truly fascinating and exceptionally humbling area. As fantastic as the human mind can be in navigating the world, we likewise make fantastic mistakes; worst of all, we are all too often entirely unaware of the errors we have made. An encyclopedic review of heuristics is outside the scope of this work, and the interested reader is referred to a number of excellent works on this issue.[21,22] However, the reader should be aware of how flawed human cognition can be.

---

[21] Gilovich T, Griffin D. 2009. *Heuristics and Biases: The Psychology of Intuitive Judgement.* Cambridge, UK: Cambridge University Press.
[22] Kahneman D, Tversky A, Slovic P (Eds.). 1982. *Judgment under Uncertainty: Heuristics & Biases.* Cambridge, UK: Cambridge University Press.

A particular example presented by Kahneman is illustrated by the following question. A bat and a ball together cost $1.10. The bat costs $1.00 more than the ball. What does the ball cost? Go ahead and answer this question in your mind. Many people will rapidly reach the answer that the ball costs 10 cents; however, this is incorrect, because $1.00 is only 90 cents more than 10 cents. Rather, the answer is that the bat costs $1.05 and the ball costs 5 cents. However, the brain can use a heuristic to change the question into the simpler construct – one that is easier to answer but gives an incorrect answer.

Of importance to this book is that heuristics will increase our violation of the deductive requirement of HD coherence, because through attribute substitution we unknowingly utilize illogical processes, or at least have changed the specifics of the reasoning construct being analyzed. Indeed, because we can substitute heuristics for the actual problem(s) at hand without being aware of having done so, we are often unaware that the heuristic exists at all. This was the brilliance of Tversky and Kahneman in discovering heuristics in the first place; how does one become aware of something that, by your nature, you don't perceive? Questions that are often asked are: How many heuristics are there? What percentage of existing heuristics have been described? and How many remain to be discovered? Clearly, by definition, this can't be known, as attribute substitution makes us unaware of the heuristics that we use. The humbling truth is that cognitive psychologists may have just scratched the surface of the human propensity for error.

There are numerous medical theories and practices that exert large influence as part of what is called "alternative medicine." Specific remedies are often offered for certain ailments due to the similarity of the remedy to the ailment. It is in this context that the "representative heuristic" comes into play. For example, in some circles it is a popular belief that foods that resemble certain body parts are specifically good for that body part. Eating avocados is alleged to promote the health of a woman's uterus, because the avocado has the shape of a uterus. The rhinoceros's horn is supposed to be an aphrodisiac as it resembles an erect male penis. This discussion is not meant to support or refute the health claims made about such foods, but only to point out how the representative heuristic commonly affects thinking. It does indeed sound quite reasonable that things that look similar to each other should be related. However, there is simply no deductive basis for this idea, and thus the representative heuristic gives the look and feel of HD coherence when none is justified. That an avocado is healthy for the uterus because they share a common shape is logically ridiculous. If you were lying in bed acutely ill with a bacterial infection and were presented with two different foods that might help restore your health, would you prefer a ripe apple beaded in moisture and bursting with freshness or a moldy chunk of decaying bread covered with spores of the *Penicillium* fungus (the natural source of penicillin)?

THE HIDDEN BENEFITS OF HEURISTICS AND COGNITIVE ERRORS

It is likely that human heuristics and errors in logical thinking have been used to inappropriately indict human cognition. These cognitive errors can be uncovered in specific laboratory settings when subjects are challenged in particular ways – typically only a percentage of people utilize the heuristic – the others do not. The discovery of heuristics and a multiplicity of cognitive biases has resulted in a kind of epiphany, rejecting the previous concepts that the human mind was a rational instrument. However, it is important to realize that our minds evolved to adapt to specific conditions we might have encountered as nomadic hominids. If you perceive a large creature running at you at high speed, it probably would not be a good idea to sit down and critically analyze if it really is a large creature and if it means you harm – it would be better to just get the heck out of the way. The heuristics and mechanisms to which humans have access can allow rapid and adaptive decisions that often may be right and give a great advantage to the individual who employs them.

Models of human cognition have now emerged in which heuristics are used in situations that require rapid reasoning, whereas more analytic reasoning is used in situations that are less urgent.[23] That such mechanisms may also lead to errors in particular situations does not mean that human cognition is flawed overall; rather, it may be excellent for the environment in which it evolved. However, humans clearly did not evolve in scientific laboratories performing controlled studies of nature. That our base cognitions may not always work in such a setting is of great consequence to the practice of science, but does not indicate that heuristics and biases are always bad. Indeed, specific study of scientific thinking and discovery have demonstrated that some cognitive errors are absolutely required in the discovery phase of scientific reasoning, so long as they can be tamed by a more analytic mind later on in the process.[24] Nevertheless, to learn how to avoid heuristics and cognitive biases in the setting of scientific reasoning is a big part of the training of scientists. It is difficult because it requires us to "unlearn" a process baked into our cognitive processes by millions of years of adaptation.

In summary, deductive coherence between causes and effects, between hypotheses and observations, is required if we are going to be able to make progress in our understanding by exploring the natural world. The deductive component is like the chain of a bicycle, linking the pedal sprocket to the wheels. Without effects being the necessary outcome of causes (at least in some way and to some degree), the bicycle's wheels simply won't turn no matter how fast one pedals. As discussed previously, these consequences can be probabilistic and applicable to populations as opposed to individuals, but the probability distribution still needs to be a predictable result of causes.

There are a variety of ways for belief constructs to lose deductive coherence. The structure of the belief construct itself may

---

[23] Kahneman D. 2011. *Thinking Fast and Slow*. New York: Farrar, Straus and Giroux.

[24] Mynatt C, Doherty ME, Tweney RD. 1977. "Confirmation Bias in a Simulated Research Environment: An Experimental Study of Scientific Inference." *The Quarterly Journal of Experimental Psychology* **29**: 85–95.

preclude a deductive component, as I have argued is the case for any system where outcomes are caused by an intelligence that encompasses a free will that can change its mind and be inconsistent. If the identical situation does not result in the same outcome because the intelligence may change its mind, then it is a nondeductive system. Later chapters will expand on this with regards to being rule-governed in general. A system can also fail to be deductive if reasoning includes errors and fallacies. This can be the result of a simple error in logic, poor reasoning due to difficulty in understanding the probability dynamics of the system (as in the Monty Hall example), or due to cognitive biases whereby our minds do not appreciate the real complexity of the problem and substitute a simpler construct that does not apply (as in the case with attribute substitution and heuristics). Indeed, there is an area of study that specifically focusses on how humans get formal syllogistic arguments right or wrong, based on conditions and how the arguments are structured.[25]

Like many of the characteristics of science we will define in this work, HD coherence is necessary for science to be carried out, but it is not alone sufficient. Systems of analysis and groups that carry them out can have complete and valid HD coherence and yet be nonscientific. We will explore additional requirements for scientific practice in subsequent chapters.

---

[25] Evans, J St BT. 2017. "Belief Bias in Deductive Reasoning." Rüdiger PF (Ed.). *Cognitive Illusions.* New York: Routledge, pp. 165–81.

5   Natural Properties of a Rule-Governed World, or Why Scientists Study Certain Types of Things and Not Others

> Science is a way of thinking much more than it is a body of knowledge.
>
> – Carl Sagan

OBSERVATION OF THE NATURAL WORLD IS THE ARBITER OF SCIENTIFIC KNOWLEDGE CLAIMS

While hypothetico-deductive (HD) coherence is required for science to be performed, it is the observable predictions of the theories that most scientists investigate; in other words, the phenomena of the natural world. Science depends upon natural phenomena as the final metric of validity. Humans are persuaded by all manner of things, many of which are emotional or authoritative in nature, and in some ways the actual practice of science is no different. However, in an ideal scientific world – the world that scientific practice strives for – the final word on "truth" is not authority, revelation, or statements of a definitive text; rather, ongoing observation of the natural world around us is the determinant of how we evaluate specific scientific facts and theories.[1] Most people recognize that scientists perform studies and experiments, which are essentially a way to "check in" with the natural world – to determine whether a theory's prediction is

---

[1] It is acknowledged that humans come from and are part of the natural world, and thus what humans think is also part of the natural world. However, the term "natural phenomena" does not typically include all of the psychological, anthropological, and sociological processes that would lead to human beliefs and cultural norms – other than within fields of science that specifically study those processes

what actually occurs. The importance of this process of checking in – of using the natural world and natural phenomena as the ultimate arbiter of legitimate knowledge claims – cannot be overestimated. Creative thinking, to be sure, is a large part of the process that leads to scientific progress. Without great creativity, novel hypotheses cannot be retroduced, innovative auxiliary hypotheses cannot be generated, and new technologies to test predictions cannot be invented; however, creative thinking and imagination are not the "scientific" part of the process. Rather, the scientific application of innovative and creative thinking is found in the abilities of new ideas or explanations to resolve current violations of HD coherence where predictions and observations are misaligned, or to give rise to new predictions of the natural world, which can then only be tested by observation or experimentation.

For example, from the view of science, the debate over whether the Earth is getting warmer will be decided over time by the compilation of accurate and precise measurements of terrestrial temperature. There may be disagreements about how the measurements are taken, how long a trend needs to be observed to be considered real, and how much of an increase is meaningful, but it is still the measurements that make the determination. Ultimately, to the scientific community, it doesn't matter if the political mouthpiece of an environmental group states the Earth is getting warmer or if the CEO of an oil company states that it isn't. The scientific issue is determined by observations of the natural world. Whether public opinion and/or policy makers accept the results of scientific process and/or choose to act on them is a different issue.

Because scientific ideas must be consistent with what has been observed in the natural world, an ever greater and expanding base of information derived from observation of the natural world is required to refine old theories or to generate new ones. Thus, in addition to focused experiments meant to check the predictions of a particular idea against the natural world, scientists also gather encyclopedic knowledge of the world to add to the base of information regarding natural phenomena. Astronomers have spent millennia counting, categorizing, and describing the different heavenly bodies in order to characterize the celestial nature that was out there, and these activities continue in the current day. Aristotle and other ancient Greek scholars spent much time just describing different classes of animals and plants, and such activity persists among naturalist biologists today. As another case, the Human Genome Project that has generated the complete DNA sequence of thousands of people (with many more underway) is a massive step forward in our understanding of the human genetic makeup. Indeed, multiple efforts to sequence numerous species and their intrinsic variability are underway. These activities may be simple observations when they are performed, but they expand and fill out the HD web of belief, in meaningful and often unanticipated ways.

For a classic and historical example of how the natural world is the arbiter of scientific fact, but not facts in some other belief constructs, let us return to the issue of our solar system. It seems like a clear human observation that the Sun rises each morning in the east, traverses the sky, and then sets in the west. Simple induction uses this information to predict that the Sun will rise tomorrow, which gives some predictive knowledge, albeit imperfect (as discussed in Chapter 1). However, one can also use this information to guess at unobserved mechanistic underpinnings that lead to this outcome (in other words, retroducing a hypothesis as per Chapter 2). One person may retroduce that the Sun is orbiting the Earth, which would clearly lead to the observed rising and setting of the Sun. In contrast, another person may retroduce that the Earth is orbiting the Sun and also rotating on its axis, which also explains the observed rising and setting of the Sun, giving the illusion of the Sun orbiting the Earth when this is not actually the case. Both explanations are equally consistent with the observation that the Sun rises and sets. Accordingly, the issue cannot be unequivocally resolved on this information alone, as both hypotheses equally predict that which is observed, leading to a state of relative equipoise between the two models. In such a case it is necessary to continue to further assess the issue in an attempt to clarify the situation.

In medieval times, Holy Scripture and the Pope declared that the Earth did not move; the Earth was the center of the heavens and the Sun orbited the Earth. Given this position, medieval theologians and scientists both generally accepted that the Earth was indeed at the center of the solar system with the Sun orbiting it. At this point, for the theologian, the debate was resolved, and no additional intellectual energy needed to be spent; a clear and infallible answer had been generated by divine providence and scripture.

In contrast to the infallibility of many religious claims, science has long ago acknowledged that scientific facts that are held as fundamental truths by one generation can be rejected by subsequent generations. Thus, an important (and essential) part of science is that it is an iterative process. The theologian has a firm answer to the question and need not consider alternatives; indeed, to do so may constitute heresy. In contrast, the scientist will consider new observations as they become available (be they methodically sought out or just noted through general observation, such as the phases of Venus) and attempt to reconcile them with current understanding. If the new knowledge is incompatible with the existing theory, then some scientists will assault the theory; however, doing so in the church would be forbidden. Thus, what is standard practice in science – the constant and never-ending scrutiny of predictions made by existing hypotheses – is strictly prohibited in certain theologies.[2]

Placing the Earth at the center of the solar system with the Sun orbiting it results in different planetary motions and phases than if the Earth is orbiting around the Sun and the Earth is also rotating around its own axis. Astronomers collected data, and their findings were inconsistent with the Sun orbiting the Earth. In this case, scientific thinking would conclude that, in fact, the Earth is not at the center of the solar system, because such a system was now understood to be

---

[2] Of course, many systems of theology have tremendous scholarship and certainly analyze the world thoughtfully and even critically, just not scientifically.

inconsistent with what one could observe in the real world; or at the very least, that it explained less of the observed world than a Sun-centered solar system. Great societal pressure may be put on the scientist to not reach such a conclusion, and societal pressures do affect scientists, but in ideal science the data are what the data are: "*Eppur si muove.*"[3] Of course, the everyday practice of science may not always live up to its ideals, but the guiding ideals are nevertheless different than authority based systems. Of course, should additional data come out later in support of an Earth-centered solar system that outweighs existing data, then the scientist should revert back to the original point of view. Perhaps more importantly, if sufficient data accumulates such that neither an Earth-centered nor a Sun-centered system can explain the findings, then both must be rejected and a novel system needs to be retroduced that can encompass the new observations. This requirement of science, to use observations of the natural world as an arbiter of knowledge claims, is fundamental.

SCIENCE IS AN ITERATIVE, CORRIGIBLE, AND SELF-CORRECTING PROCESS BASED ON FALLIBILITY

In science, all truths are tentative and subject to ongoing challenges based on new observations and experimentation. The priests who believed in an Earth-centered universe found themselves in a very different mindset. The Bible and the Pope had declared a certain conclusion to be correct, and this determination was not susceptible to further examination or rejection – at least not in public or as a process of the religion. Even if others went out and collected additional data that were not consistent with the Pope's edict, no amount of data could lead to a rejection of the Pope's declaration, which was

---

[3] Galileo Galilei is reported to have muttered these words under his breath after being forced by the Pope (under threat of imprisonment and torture by the Holy Inquisition) to recant his heretical notion that the Earth moved around the Sun and not the other way around. It is translated as "and yet it moves" in reference to the Earth moving, which is in disagreement with biblical interpretation. It is unclear whether he ever uttered these words, and it is likely an apocryphal story, but is often used by scholars of science to illustrate a point.

infallible and absolute.[4] Of course, even priests are human, and it is natural to question and doubt, but even if one's faith did waiver it was not acceptable to express it and certainly not in an official capacity. Indeed, one of the greatest virtues of faith is that it is an unwavering belief despite experience and not because of it; this aspect is one of the major disconnects between religious and scientific thinking.

It is necessary to note that later Popes have indeed overturned previous Papal conclusions. Thus, even faith-based theology can change and adapt, but it is not the main focus or goal of the theological process to seek change and to challenge doctrine based on observation of the natural world. Rather, the preferred modality is to seek and find ways to reinterpret new data such that infallible doctrine remains intact, and a great number of intellectual giants of theology have devoted their lives to this task. Again, this is a fundamental difference between science and faith-based systems. In science, ever more data is to be accumulated, and all premises and conclusions are always suspect. By definition, faith is a belief in a principle or idea that not only doesn't need data to support it, but that persists despite a great deal of data to the contrary. Science takes precisely the opposite point of view. This is not to say that many religions don't encourage critical thinking; they do, and it may even involve the questioning of God. However, this is a maneuver to better understand God, not to test God's existence.

From a certain point of view, essentially all human systems of belief are structures seeking coherence or agreement between ideas (except in the case of people suffering psychosis, perhaps). In other words, all systems are webs of belief between multiple ideas and observations. This is one reason why it is so difficult to find clear demarcations between science and nonscience, as they have so much in common – humans seeking coherence between beliefs and

---

[4] There is no intent to single out the Pope in an unfair way. The argument holds for all authorities, religious and otherwise, and the Pope is used here because of the great historical example of Galileo and the solar system.

experience. However, the rules by which one is permitted to modify the web of belief differs and can be a defining characteristic. Priests are not allowed to modify the web of religious belief by attacking the base premise of God. Scientists are not allowed to modify the web of belief in deference to authority over experience and must modify it based on evidence. There are acceptable maneuvers in science by which one can discount an observation (e.g., it was a chance occurrence and not a real effect, as explored in detail in Chapter 9; or, it was improperly detected or interpreted); however, ignoring an effect because an authority tells you to do so or because you don't like the answer and its implications is not acceptable.

This is not meant to imply that individual scientists or groups of scientists may not be dogmatic (they often are), nor is it meant to imply that scientific paradigms and the scientists who support them don't ridicule new ideas or act as bullying authorities (this also occurs). For example, when the great immunologist Louis Pillemer described a novel mechanism by which the immune system fights infection, one that was contrary to existing scientific dogma, he was so ridiculed and discredited by the scientific establishment that he committed suicide.[5] However, whereas authority is the stated ideal in many faith-based systems, such is not the case in science. It is precisely for this reason that other scientists kept observing and checking in with the natural world, and 10 years after Pillemer's death, numerous scientists discovered that he had been right and his ideas were heralded as genius. Scientific process remedied the issue (from the standpoint of knowledge, if not personal life), although it was clearly a process with a tragic outcome. When authority serves as a basis for knowledge in science, it is not doing so by design. Because new observations and ideas can challenge multiple existing parts of the web of belief, and thus may have to address substantial evidentiary

---

[5] Lepow IH. 1980. "Presidential Address to American Association of Immunologists in Anaheim, California, April 16, 1980. Louis Pillemer, Properdin, and Scientific Discovery." *Journal of Immunology* **125**(2): 471–5.

weight against them, it is good and appropriate for scientists to insist upon exceptional evidence to back up exceptional claims. In this way, science can be fairly conservative and resistant to change. However, this is not the same as ignoring and mocking a result, and ridiculing the scientist who made it because one doesn't like the result and can use authority to indict it. Rather, empirical claims should be followed up with empirical investigation, with the natural world (and not authority) as the ultimate arbiter.

Just because someone is a professional scientist doesn't mean that they will act scientifically in all cases and at all times (even scientists are human, after all). Moreover, societies of scientists are still prone to the madness of crowds and groupthink problems, which is a regrettable trait of humans interactions. However, when societies of professional scientists act in this way they are not acting scientifically; rather, the scientific practice is to attack and modify its previous interpretations as new information about the natural world is generated. When scientists act as though no notion is above rejection or modification if it ultimately fails to explain the natural world, they are acting within these traditions. Likewise, just because someone is highly religious or a professional clergyperson does not mean they will never question authority or even the word of God. However, when this person does so, he or she is not acting in the tradition of the theologian (in many faiths), who uses basic and sacred tenets of their belief and faith to guide how they process and interpret their experience. For the professional scientist, the opposite is true[6].

THE ABILITY TO CHECK AND RECHECK NATURAL PHENOMENA

The difference between science and nonscience is also found in the nature of their respective knowledge bases. A young student of

---

[6] Indeed, many of my closest friends are other scientists who actively attempt to discredit my ideas in public forums. This is not offensive or unusual in anyway, it is one of the norms of the scientific culture.

theology and a young student of biology will initially receive similar types of indoctrination into their chosen area of scholarship. The reader should make no mistake: Students of science are no less indoctrinated than are students of religion.[7] The scientist and the theologian will each read extensive texts containing large quantities of facts about their fields. Likewise, both will listen to extensive lectures by faculty who are passing their expertise to the next generation. In each of these cases, the legitimacy of the truths being taught relies on trusting the information being dispensed, based on the authority of the person dispensing it. The science textbook and the holy text both contain extensive factual information, which students typically accept to be true. Likewise, students typically accept the authority of professors and the validity of the information being provided. At this point there is no difference whatsoever between the student of science and the student of theology, other than the subject matter being studied. They have each been told facts and have read books on their subjects. The lectures are just spoken words from a person, and the book is just printed words on paper; after all, anything can be said, and anything can be written. At this point, in both cases, it is purely an issue of trusting authority.

A central tenet of religion is that the theology student takes it on faith that the teachings and the Holy Scripture are true. This is not to say that theology students do not scrutinize or question what they are learning and explore the details with great academic acumen; however, the base assumption is that the written and spoken words are correct and meaningful, if not literal. If one still cannot reconcile what is taught with experience, or if fundamental questions arise, then eventually, faith can be evoked as an instrument to remedy any discord of ideas. One may seek additional meaning or alternate

---

[7] I do not wish to encourage the idea that science and religion are opposed to each other in intent, which I do not believe needs be the case. Rather, religion is chosen as an example, because it seems to be a generally agreed upon class of belief constructs that is not a science. The middle ground of pseudosciences is focused on later in the discussion.

interpretations in the words, but the fundamental principles should not be changed. This lack of agreement of ideas or experience can be accepted, as a mortal cannot (and maybe should not) understand the unknown universal plan of a god or gods. It is accepted as a matter of faith, and such acceptance is typically a virtue.

It is essential to appreciate that many science students are sadly mistaken early in their training when they state that science is not faith based. At this point, the science student is identical to the theology student, in that the validity of the scientific knowledge being taught is taken completely on faith. The student has been told something and has accepted it as true without any personal experience in the matter one way or the other. Yet a fundamental difference still exists between science and religion. It isn't the ability to generate new knowledge, for just as the science student can do research and pursue new knowledge in the context of existing science, so can the theologian do theological research or gain new religious experiences in the real world.

As is very likely in life, both individuals will at some point encounter new information that may not obviously fit with what they have learned. The theologian may encounter experiences that don't make initial sense in the context of his or her faith in the teachings (thus breaking coherence). Likewise, the scientist may perform studies that seem inconsistent with the scientific knowledge that he or she has been taught. However, it is at this point that a fundamental and essential difference emerges. The scientist can go back and repeat almost any study that has been described within the base knowledge upon which new knowledge is being developed.

It is important to point out that written scientific knowledge and religious knowledge are both historical. The difference is that a biblical scholar can't go back in time and see the Flood, whereas I can repeat an experiment that someone else has done in the past. Precise methodologies may be obscure, some materials or equipment may not be available, and one may always be limited by the resources one has to carry out experiments; however, there is nevertheless a way by which even the most fundamental pillars of scientific knowledge (upon which so much rests) can be retested and reevaluated. If new data seem inconsistent with DNA having a double helix structure, then the studies of Franklin, Watson, and Crick can be repeated and reexamined. However, if the theologian has experiences that seem to question stories or tenets from a holy book, nothing can be done except attempting to reinterpret the new experience in the context of what the book says. People can seek additional texts and can even look for archaeological evidence of biblical events; however, the theologian cannot recreate Sodom and Gomorrah and witness their destruction.

It is precisely because one can repeat most experiments upon which scientific knowledge rests that scientific facts are never safe and are always under assault from new developments, new experimental tools, and new ideas. In practice, this may differ in the case of biology compared to physics and chemistry, as it is assumed that atoms and simple chemical compounds behave the same way now as they did hundreds or thousands of years ago and that basic laws of physics have not changed.[8] In contrast, biological populations do change over time.[9] The same strains of germs may not exist, and thus it may not be possible to repeat the exact studies of Louis Pasteur; however, enough related science can be carried out to recheck most previous claims. Science is not only iterative but is corrigible as well, and can be revised over and over again. Other systems of knowledge that are not susceptible to fundamental reexamination of the pillars upon which they are built simply cannot be subjected to the same type of rigorous reassessment. In such fields of study, mistakes,

---

[8] The problems of induction make this assumption not based in logic, and even laws of physics may theoretically change over time.

[9] It should be noted that how such change may reflect on where diversity of species came from is relevant only to Darwin's ideas regarding the origin of species, which occurred before we have recorded scientific observations. However, the changing of species since humans have been observing is clear and uncontroversial. This is explored further in subsequent sections.

misobservations, and misinterpretations of the past cannot be tested and remedied moving forward, at least not to the same extent as in science.

THE DATA ARE WHAT THE DATA ARE, AND IF YOU OBSERVED IT, YOU OBSERVED IT

In science, the data are sacrosanct. One can question whether the data were collected accurately and/or interpreted correctly (e.g., challenge the observation part of HD coherence); however, the data may not be purposefully altered, excluded, or fabricated simply because they don't align with what is predicted. Because observation of the natural world is the ultimate metric, purposefully altering such observations violates the most important canons of scientific ethics and constitutes professional misconduct that can prohibit funding for one's research, prevent publication of one's findings, and end one's scientific career. Once they have been collected, the data are what the data are. If the data reject one's favorite hypothesis, then that's just too bad. In the poetic words of Thomas Huxley, the great tragedy of science is "the slaying of a beautiful hypothesis by an ugly fact."[10] At the end of the day, it doesn't matter how appealing or elegant an idea is or how much one wants it to be true; if the data are inconsistent with that idea, the data cannot just be altered. This is, of course, not to say that the data can't be questioned, reproduced, and scrutinized; indeed, this is compulsory for good scientific practice. However, the data cannot be changed just because they break coherence, and if data are challenged it must be done so according to scientific rules for updating the web of belief, as justified throughout this text and described in Chapter 13.

The sacrosanct nature of data is by no means a common notion outside of science. In law, for example, at least in the context of American jurisprudence, data are not sacred and immutable. Consider

---

[10] Huxley T. 1870. "Biogenesis and Abiogenesis." Presidential Address at the British Association for the Advancement of Science.

a citizen charged with murder. Both the prosecution and the defense are required to disclose their findings to each other, meaning that unless something has gone wrong, they have the same (or at least similar) data – in this case called evidence. The unequivocal hypothesis of the prosecutor is that the accused is guilty, whereas the defense is just as devoted to the hypothesis that the accused is innocent (or, at the very least, that the state cannot make the case that the accused is guilty). In the case of defense attorneys, if the evidence doesn't support the hypothesis (i.e., my client is innocent), then they have it within their discretion to advise their client to plead guilty and seek a lesser sentence. In all likelihood, such advice will not depend upon whether the client is guilty or not, but rather on the likelihood of winning at trial what plea deal can be negotiated, and the potential sentence or penalty of losing.

However, if the client will only accept an innocent plea, then the attorney has a professional obligation to do whatever he or she can (within the law) to exclude damning evidence from being heard in the courtroom. Should the attorney fail to exclude the evidence, he or she will attempt to discredit, twist, and warp the evidence to maintain and support the hypothesis that the accused is innocent. There are limits on how data can be altered, to be sure, as the attorney cannot knowingly suborn perjury from a witness, but by calling on the accused to testify in the narrative they can allow evidence that is "known" to be untrue to enter the record. While the defense attorney cannot intentionally fabricate data by introducing evidence he or she knows to be false, the defense attorney can certainly exclude and attempt to discredit prosecutorial evidence known by the defense to be true. Under essentially no circumstances whatsoever will defense attorneys alter or adjust their main hypothesis (innocence of the client) to fit the data. Of course, in the case of legal defense, this is not only appropriate but also imperative. Imagine what would happen if, after considering the evidence, the defense attorney became convinced of the client's guilt and then adjusted the hypothesis to fit the data. The attorney would then stand up in court and proclaim her or his belief that the client is in fact guilty and ought to be convicted of the crime. This would be a profound breach of legal ethics and constitute malpractice, potentially leading to disbarment of the attorney and a mistrial. In other words, loss of HD coherence (the data don't support the hypothesis) cannot be remedied by altering the hypothesis, only by challenging the data or background assumptions, and the data and background assumptions can be twisted by attorneys in ways known to the attorney and/or the accused to be incorrect in order to save the hypothesis.

The prosecution does have it within their discretion to change their hypothesis (in this case, to drop the charges against the accused). In these cases the prosecution typically will simply drop the charges rather than declaring or arguing for the accused's innocence, although admittedly proclaiming innocence is not typically the prosecutor's job or appropriate role in the process. Regrettably, due to myopic zealotries or political ambition, there have been numerous documented cases of the prosecuting attorney's office purposefully withholding exculpatory evidence from the defense and, in some particularity egregious cases, manufacturing false evidence. Thankfully, such cases are rare (or at least hopefully so); most prosecutors are likely good and honorable people, but it seems it is not a compulsory part of their job to abandon a hypothesis, even when the data clearly reject it.

Political processes seem quite similar (at least in the United States). The hypothesis of any given campaign is that their candidate is the best person for the job, or from a more cynical point of view, either the most electable or the easiest to control. Nevertheless, the apparent working hypothesis put forth is that they are the best candidate. If data exist or come to light that argue to the contrary, the data will be attacked, excluded, perverted, and minced into a thousand pieces to avoid altering the hypothesis. Advertising data arguing that one's opponent is a poor candidate are also a large part of campaigning, and it is worth noting that sending out investigators to discover unsavory facts about one's opponent and then informing the public of accurate facts, is consistent with the finest traditions of scientific practice (i.e., using data as an arbiter). However, the purposeful misrepresentation of data with the intent of impugning one's opponent, either through rumor or direct intimation, is regrettably also a standard maneuver in the political game and anathema to scientific practice.

At the time of this writing, the previous presidential election in the United States was an incredibly contentious and controversial event. Both sides had accused the other of outright lying and deceit, with both sides also denying having done so themselves. The very basis of evidence had been undermined, with the term "fake news" being applied to things that may (or may not) be true, but which those using the term simply did not like. This rancor is now extending itself throughout American politics. Although the pitch of rhetoric may be amplified above normal, this is certainly nothing new. If we simply go back one electoral cycle, a striking admission of the permissibility of the intent to mislead occurred during the 2012 Republican presidential primary. In the 2008 presidential race, Barack Obama criticized his opponent John McCain with the following quote: "Senator McCain's campaign actually said, and I quote, 'If we keep talking about the economy, we're going to lose.'" In 2012, Mitt Romney's campaign altered the data by removing the reference to Senator McCain's campaign, taking President Obama's quotation out of context, and showing a clip of Obama saying, "If we keep talking about the economy, we're going to lose." This left the impression that Obama was stating his own view and not McCain's. When the Obama campaign and the media confronted Romney's campaign, a senior Romney advisor (Tom Rath) told CBS news "that was his [Obama's] voice...he did say those words." In reference to this issue, Romney himself said, "What's sauce for the goose is now sauce for the gander."[11] This is a brazen admission of how the purposeful misrepresentation of data is common in politics; the Romney campaign is in

---

[11] Zeleny J. Nov. 24, 2011. "Romney Defends Ad Aimed at Obama." *The New York Times*. https://archive.nytimes.com/query.nytimes.com/gst/fullpage-9F03E3DC1F31F937A15752C1A9679D8B63.html

no way unique in these activities. While some campaigns indulge in such skulduggery more than others, distortions of this type are ubiquitous in both parties and amongst independents as well. It is an easy study of history to find that such maneuvers and practices have been present since the beginning of the Republic and far earlier as well.

As in a legal trial, it's not an option to alter the hypothesis that one's candidate is the best person for the job. One could certainly argue that the hypothesis is altered when a candidate withdraws from a race, although this is more likely from a determination that the candidate can't win rather than an admission that the candidate is flawed. Nevertheless, it is not only fair but good political savvy to manipulate or hide data in the face of a sacred hypothesis, and in this regard the basis of political truth is at odds with the basis of scientific truth.

These arguments are in no way intended to imply that professional scientists don't sometimes alter data; they clearly do, and some have been caught doing so. Rather, the point is that while altering data with the intent to deceive is forbidden in science, the standards are much looser in other areas. Sometimes the "flexibility of data" is present for laudable and ethical reasons. American law does not allow the admission of evidence into court that is improperly obtained (e.g., without a warrant), and this is a highly ethical position weighing the general rights of society against unreasonable search and seizure over the particular specifics of one case. Nevertheless, this serves as a point of distinction between science and other areas, as science has no mechanisms or circumstances that justify ignoring data that is believed to be correct.[12] Thus, the allowable norms of treating data

---

[12] It should be noted that in some extreme cases where scientific data are obtained unethically, some may argue that the data should not be used even if it would change scientific knowledge, as doing so would validate an unethical practice, but this argument is unclear and controversial. Others would argue the data should be used such that those who suffered from the breach in ethics should not have done so in vain, but that this in no way would justify future unethical behavior.

can help to distinguish some approaches from others. Although using these general criteria partially defines science as a thinking modality and can exclude other systems for which data are not sacrosanct, such does not mean that all thinking modalities for which data are sacrosanct fall within the definition of science. In other words, much like HD coherence, the sanctity of data could be argued to be a necessary condition for science, but not a sufficient condition, because there are many thought constructs that do not allow for the alteration of data and yet which may not be categorized as science.

In summary, this chapter focusses on the sanctity of data and the flexibility of other parts of the web of belief in science, in contrast to the flexibility of data and the sanctity of other parts of the web of belief in many non-sciences. Indeed, whether or not a system of thought can alter its views based on the data is extremely important for understanding the nature of science. In the notable words of the famous physicist Carl Sagan, "In science it often happens that scientists say, 'You know that's a really good argument; my position is mistaken,' and then they would actually change their minds and you never hear that old view from them again...It doesn't happen as often as it should, because scientists are human and change is sometimes painful. But it happens every day. I cannot recall the last time something like that happened in politics or religion."

THE NEED FOR REPRODUCIBILITY AND REPETITION: A REQUIREMENT FOR A CONSISTENTLY RULE-GOVERNED NATURE

One of the things scientists talk about with great frequency is the reproducibility of their observations. In other words, if one researcher performs an experiment and observes a result, can a similar result be observed when the experiment is repeated over and over again? Moreover, can a scientist in a lab halfway across the world observe the same thing? Concerns over a lack of reproducibility have become a major focus in recent years, as it has become clear that many experimental results can't be duplicated by other scientists.[13] This has led to all manner of turmoil. Why has this issue of reproducibility caused such an uproar? If a spiritual person takes a walk on a wooded path, feels the presence of energy in his or her being, and tells other spiritualists about it, no one will tell the spiritual person that it wasn't real because other people didn't have the same experience when they walked down the same wooded path. If anything, other people not having the same experience may make the experience even more profound, as it becomes a personal connection between the spirit and the person. In contrast, there are several reasons why reproducibility is something that scientists focus on so intensely and why it contributes to defining what science is and how it works.

First, the issue of reproducibility speaks to fluke errors of chance that appear to, but do not in actuality, reflect natural phenomena. In an effort to draw general conclusions from specific observations of nature (i.e., to make inductions as explained in Chapter 1), there is always a concern that one has not actually observed a true aspect of nature but has been fooled by random noise in the system (this is explored in more depth in Chapters 7 and 9). However, for the current discussion, it is important to note that if a phenomenon occurs over and over again, then it is much less likely to be a random event that happened by chance alone. Hence, reproducibility protects against what are called "type I errors," which result from concluding some effect exists in nature when it really doesn't.

The second essential component of reproducibility, especially when speaking about different scientists and labs, is the issue of generalizability. Something may happen over and over again in one scientist's lab but may not happen in any other lab in the world. If this is the case, then scientists from around the world may visit the reporting lab and observe the phenomenon for themselves. In many well-known cases of this situation, the lab making the observation

[13] Baker M. 2016. "1,500 Scientists Lift the Lid on Reproducibility." *Nature* **533**(7604): 452–4. doi:10.1038/533452a

was making an error, being sloppy, or misinterpreting nature. However, in other cases, there was something particular about the technique in the reporting lab that was not communicated properly to other labs. Alternatively, there may have been something particular to the environment in the reporting lab that was absent from other labs. Ferreting out the difference that makes the phenomenon work not only allows labs around the world to observe and study the same thing, but can also provide key insights into mechanisms of the phenomenon. In other words, if you figure out that a particular water contaminant is required for the phenomenon to take place (present in one lab but absent in the other), then you have learned something essential about the process that you might not have otherwise known – whatever the contaminant is will help you figure out the mechanisms of what is being studied.

A third issue is that reproducibility is essential in the search for mechanistic understanding and causal associations. To illustrate why, consider famous events that have occurred in human history, such as the fall of the Roman Empire or the Great Depression. Countless hours of energy and thought have gone into the analysis of why the Roman Empire fell and why the Great Depression occurred. Many scholars have analyzed these questions and retroduced a wide variety of different hypotheses, each of which predicts the data (or much of it) and many of which are mutually incompatible with each other. As is the practice in science, historians likewise ask themselves: How is one to assess which hypotheses are the most likely?

In historical studies, one is analyzing an event that happened in the past and will never happen again. Future empires may fall and economic depressions will almost certainly happen again, but the Roman Empire will never exist again, and the precise geopolitical and economic conditions present in 1929 will never occur again. How, then, is one to distinguish between different explanations? Certainly, historical studies can gather data that is meaningful in the form of records, historical documents, correspondence, etc.; such historical data can provide substantial evidence for or against explanations. However, that is the limit of the analysis one can do. Suppose one has the hypothesis that the Roman Empire fell due to the use of lead in dishes that resulted in the poisoning of Roman leaders and a decline in their ability to govern. Unless one can invent a time machine, go back to ancient Rome, prevent lead dishes from ever being used and see if the empire doesn't fall in the same way, one can never test the hypothesis directly. Clearly, no one has yet invented a time machine, or if they have, they have kept it to themselves (now, in the future, and in any past they visit).

This is precisely the problem that is remedied by a system that is robustly reproducible. If a phenomenon happens over and over again in a similar way each time, then one is essentially using a time machine. If a scientist has a highly reproducible system, then she or he can remove factor A and see if the phenomenon still occurs. If removing factor A has no effect, then she or he can conclude that factor A is not required. If removing factor A prevents the phenomenon, she or he can conclude that factor A is necessary.[14] This mechanistic knowledge would then extend not only to previous iterations of the phenomenon but also to future iterations. This assumption of exact reproducibility of the phenomenon over time is a big assumption, and once again steps into the problems of induction that were raised in Chapter 1. However, the problems of induction are something we have to live with, and science makes its progress in this context.

Consider a video game you have enjoyed playing, which was difficult at first. Initially, such a game is essentially a puzzle, as you don't yet know the rules and haven't mastered the skills needed to win. Over time you learn more and more about the rules and strategies, and get better at the game. Your ability to improve depends upon playing the game over and over again, seeing what works and

---

[14] This type of clean logic is often used in the description of science, but regrettably, the real process of updating the web of belief is much muddier than this, allowing auxiliary hypotheses and alternate interpretation, as is described in Chapter 3 and later in the book.

what doesn't, and modifying your strategy and/or playing techniques based on what works (and avoiding what doesn't work). In a way, this is like going back in time to encounter the same event over and over again, so that you can test what will lead to your preferred conclusion (i.e., you winning the game). Now, consider what would happen if the rules of the game and the approaches that are successful changed every time you played. Moreover, what if the rules changed at random, so there was no pattern whatsoever as to how the game could be played? In this case you would no longer get better at the game each time you play; indeed, in such a scenario your past experience with the game may make you a worse player because your strategy is being guided by information that may no longer be relevant or useful. In such a case there can be no progress in getting better at the game or improving your chances of winning other than by random chance alone. Thus, consistency over time (or, in other words, a controlled reproducibility) is essential to any increase in knowledge with regards to predicting or controlling. It is precisely for this reason that reproducibility of systems is so essential to all of science and why scientists put such a premium on such issues. Without reproducible systems, there can be no purposeful forward progress through direct experimentation in the system[15]. This issue wanders again into the area of induction, as there is no logical basis to conclude that the universe will behave the same way tomorrow as it does today; but this is a risk that we have to live with and always be aware of. On the flip side,

---

[15] This should not interpreted that no science can be done on "one-time" events. However, the nature of the scientific study will be different, and limited in its ability to perform direct mechanistic studies. If there is a web of belief that intersects with the one-time event, then theory can mature, guided by direct experimentation in parts of the web that are accessible and reproducible, and data of the one-time event can be used to guide theory. In other words, would our best current understanding predict that which was observed, even though it was only a one-time observation. As an example that is often used, supernovas are things we observe at a great distance and very rarely; however, the web of belief of astrophysics intersects with other parts of physics. If physics theory develops in a way that does not predict data we have observed about the few supernovas we have seen, this can constrain the development of the web in a scientific way.

unless the behavior of the universe tomorrow is at least somewhat related to how it behaves today, induction will not work, and science will be unable to predict the unobserved. If the rules of the universe changed at random, there could be no science at all.

THE SINE QUA NON OF SCIENCE: A RULE-GOVERNED UNIVERSE

Is it true, as has been argued, that science is different in fundamental ways from other knowledge-based systems because it has no sacrosanct premises that cannot be rejected? Is it a correct statement that science will attack any idea, no matter how fundamental, if that is where the data leads? In my view the answer is yes, but there is one exception. For science to exist, the universe must be a rule-governed place, and if the rules change, then there must be at least some pattern to the change. If the rules of the natural world change at random, then one cannot predict or control anything, and one cannot gain any knowledge of the unobserved world based on the observed world. A rule-based universe that is not completely random is the bedrock premise of science. In the words of Daniel Dennett, "No rational creature would be able to do without unexamined, sacred things."[16] A rational, rule-governed nature is sacred to science. It is for this reason that the advancement of quantum theory in the last century was so distasteful to many scientists, as it suggests an intrinsic randomness to the universe. Such is not practically the case, as this is not how we encounter the macroworld, and even probability distributions are rule governed (although only applicable to populations of things but not individual things, and as such, seemingly not deterministic when attempting to predict single instances).

Nevertheless, revulsion at the concept of randomness in nature is the basis of Albert Einstein's famous quote, "God does not play dice with

---

[16] Wolf G. Nov. 1, 2006. "The Church of the Non-Believers." *Wired Magazine.* www.wired.com/2006/11/atheism/

the universe." This comment reflects the fundamental belief that a rule-governed universe is the sine qua non of modern scientific thought.

The requirement of a rule-governed universe is a basis for excluding "supernatural phenomena" from that which science can study. The reason that supernatural cognitions are not consistent with scientific study (see Chapter 4) is that they violate the need for deducible consequences from a given web of belief because they can simply change their minds for no reason. In other words, they are not rule governed; however, what about supernatural things that are not a cognition? Regrettably, they suffer from the same problem, but not because they can capriciously change their minds – they have no minds. Rather, the very definition of a supernatural thing is that it transcends the laws of nature.[17] That is to say, for supernatural things, the rules of the natural world don't apply – they are outside of nature. Thus, supernatural things are not part of our rule-governed universe, and as such, cannot be the subject of scientific work.

This does not mean that science cannot study supernatural claims. Indeed, over time, many phenomena previously felt to be supernatural have been tackled by science by uncovering how they actually fit into the rules of nature, but then they cease to be viewed as supernatural. Similarly, specific supernatural claims can be tested by scientific methods (e.g., there have been randomized controlled blinded clinical trials on the healing effects of prayer). In such cases, should a robust and reproducible phenomena be found (e.g., on average, and all other things being equal, if patients who were being prayed for did better than those who were not being prayed for at a frequency greater than can be explained by chance alone), then one would be essentially defining a rule, allowing something that could be studied further.[18] The effect would not prove that the posited

---

[17] It should be noted that there is lack of uniform consensus on what exactly the term "supernatural" means. I am defining it as "outside of nature," and as such, not susceptible to the rules of nature.

[18] While some poorly run trials have suggested an effect of prayer, rigorous studies found no significant effect of prayer on medical outcomes. Indeed, in some studies,

cause was what was really resulting in the effect (due to retroduction and error of affirming the consequent described earlier),
but it would allow subsequent investigation. Thus, "real" effects in
the natural world that are claimed to be of a supernatural origin
can be investigated by science. However, unless and until a rule is
defined that can be observed as a reproducible phenomenon within
our natural world, no scientific progress can be made. If the
believers in the supernatural claim continue to insist on the effect
being real despite the absence of any detectable effect by methods
known to address common human errors, then they have left the
realm of science (see discussion of astrology as a specific case in
Chapter 13).

Of course, it may be that things considered supernatural simply
have a different set of rules than what we commonly think of as
nature, and were we able to define such rules, then science could
study them (they would just function in a part of nature where the
rules were altered but would still be natural). However, defining
something as simply not being part of a rule-governed system makes
it immune to the methods and approaches that science employs.

This does not mean that supernatural things cannot exist – their
certainly could be entities that are unconstrained by natural rules (or
any rules) with regards to how they exist and function, although it is
not clear what existence would mean in this case. However, this
simply cannot be studied by science. With no rules, nothing can be
predicted or controlled, and the study of it is useless with regards to
the goals of science. As such, scientists typically ignore these notions.
Whatever research programs may exist from time to time that study
things labeled as "paranormal" arc rcally an attempt to define some
kind of rule of the phenomenon, any kind of rule, so that scientific
study can commence. If no rule can be established, the phenomenon

the groups being prayed for did worse. Overall, the outcome is exactly what one
would predict if there were no effect and small differences in either direction were
the "noise" of chance occurrences.

fades into the waste bin of things that can't be studied, and likely don't exist from a scientific point of view.[19]

In this chapter we have explored the role that data play in testing scientific HD thought constructs and how this differs from the treatment of data by other nonscientific modalities. Of course, such a view presupposes that we have the capacity to correctly observe and study nature, for if we cannot obtain data in a meaningful way, then updating our belief constructs based on data will not be a fruitful venture. This chapter has raised some issues regarding the specifics of how science handles data that disagree with theory, and about the scientific rules for updating the web of belief. But before we tackle these issues in more detail in the final section of the book, in the next chapter we need to explore how well humans can actually observe the natural world. For if we cannot observe nature, then at least in science, we are lost.

---

[19] For a more detailed analysis of this issue, the reader is directed to Fales E. 2013. "Is a Science of the Supernatural Possible?" In Pigliucci M, Boudry M (Eds.). *Philosophy of Pseudoscience: Reconsidering the Demarcation Problem.* pp. 247–62. Chicago and London: University of Chicago Press

6   How Human Observation of the Natural World Can Differ from What the World Really Is

Our view of reality is conditioned by our position in space and time – not by our personalities, as we like to think. Thus every interpretation of reality is based upon a unique position. Two paces east or west and the whole picture is changed.

– Lawrence Durrell, *Balthazar*

Traditionally, scientists and philosophers of science have worked under the assumption that humans are pretty good at making observations of the natural world. Many thinkers, as far back as antiquity, recognized that experience could lead us astray and thus favored deductive systems of reasoning; however, to justify deduction, early philosophers argued for humans' innate ability to perceive fundamental truths and correct base axioms. Empiricists clearly rejected this idea, favoring our ability to observe nature by using our senses over some perception of fundamental truths. However, both camps seemed to accept that humans could observe, or at least gather base information, about the natural world in a meaningful way, although there has not been uniform agreement on this.[1]

The importance of the assumption that humans are good at observing nature cannot be exaggerated. If observations of nature are the arbiters of scientific knowledge claims, then we are cursed from the start if we do not observe nature correctly, and the worse we are at

---

[1] In addition to previously discussed concerns about whether induction is rational, there has been much debate by philosophers as to what our senses actually detect versus what kinds of observations are higher orders of processed information. There has also been debate of a more metaphysical nature over whether we can have any confidence that our observations necessarily reflect an external reality.

observing what is in front of us, the worse off we are. Most of us have high confidence in what we ourselves observe, the accuracy of our perceptions, and the fidelity of our memories; scientists and philosophers of science have traditionally been no different in their levels of confidence.

It is only relatively recently that scholars have begun to formally explore the question of whether or not humans are skilled at observing nature. Arguably, scientists have been too busy trying to observe nature to pursue a detailed exploration of whether or not they could observe nature. Moreover, the tremendous strides made by science seem to justify the sense that humans observe nature pretty well; how else does one explain our theoretical and technological breakthroughs? However, specific groups of cognitive psychologists and neurobiologists have made substantial progress in assessing the act of observation itself.[2] Based on their research, it turns out that we are not so good at observing as we previously thought ourselves to be. Even worse, we seem to be consistently overconfident about our abilities. If we adjust our theories and background assumptions to maintain coherence with what we experience, and our experience is essentially the accumulation of our observations, then any failure to observe correctly threatens the accuracy of our coherence at its very core. This is a matter of great concern and the focus of this chapter.

HUMAN OBSERVATION AND PERCEPTION: A CONSTRUCTED EXPERIENCE

There is some paradoxical irony to the notion that we use the human mind to assess the flawed human mind, but what other choice do we have? In the words of comedian Emo Phillips, "I used to think that the brain was the most wonderful organ in my body. Then I realized who was telling me this." The flaws of human observation, perception, and interpretation have been noticed for some time. Law enforcement has

---

[2] Sociologists, anthropologists, and historians have also analyzed other biases that affect observation, as is discussed in other sections.

long understood that a dozen eyewitnesses to the same event may tell drastically different stories – they can't all be right, and in some cases all of them have turned out to be wrong. While much on this subject remains to be learned, seminal controlled studies and analysis of human observation have generated some very disconcerting results, questioning the quality of human observation in general.

The five established human senses are sight, taste, touch, sound, and smell. Although the human brain likely remains the most sophisticated computer on Earth, at least for the time being, it comes with substantial weaknesses. One of the principle weaknesses is the brain's inability to process all of the data being delivered by the five senses. Indeed, the amount of information delivered by our senses to the brain is staggering. In addition, our brain has the added task of interpreting the input in real time. There are a multitude of objects in most visual fields, and they need to be characterized, categorized, and assessed as they are encountered. For example, when walking across a street, it is essential that the brain not only process visual images into recognized objects, but it must distinguish all different manner of objects. It must determine the direction, speed, and potential acceleration of these objects. The strategy for crossing a street is very different if the objects are stationary cars vs. moving cars or cars moving toward the observer vs. cars moving away. This is greatly complicated by the fact that while cars tend to have common features, they can each look and sound very different from each other, as can every object in a visual field. It is truly remarkable that our brains do as good a job as they do; of course, when they fail (as they still do with some frequency), disaster can ensue.

How do our minds handle the processing of information? To begin with, human brains are basically pattern recognition machines; they take in information and look for recognizable patterns (e.g., cars, trees). These activities extend to all the senses, including recognizing different sounds and words, being able to distinguish an object by touch, identifying a food by smell and/or taste. There are a number of different psychological theories regarding how our brains process

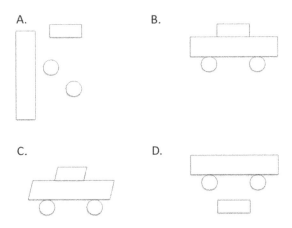

FIGURE 6.1 Pattern recognition of the human mind

raw data to generate pattern recognition, but the phenomenon is clearly demonstrable in humans and seems to be common in other animals. In the process of pattern recognition, humans assign meaning to basic input and stimuli. For example, Figure 6.1 shows two rectangles and two circles that simply represent geometric shapes (Figure 6.1A). However, putting the same shapes together in a different format generates the image of a car (Figure 6.1B). By simply tilting the boxes, the car seems to be traveling in one direction (Figure 6.1C). By moving the top rectangle to the bottom, the car becomes a face (Figure 6.1D). Indeed, the shape of a human face is hard to get away from, and one could easily view Figure 6.1B as a face with a hat but no mouth. Recognition of a face is among the first cognitive functions that can be clearly observed in human babies.[3]

This type of recognition can be seen in a second example, in which the same three circles and a single line are represented in Figures 6.2A and 6.2B, but in the former they simply represent geometric shapes, whereas the brain quickly assigns the identity of "a face" to the latter. By curving the line, one can imbue the emotion of

[3] Farzin F, Hou C, Norcia AM. 2012. "Piecing It Together: Infants' Neural Responses to Face and Object Structure." *Journal of Vision* **12**(13): 6.

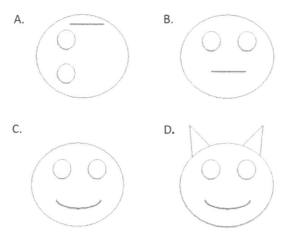

FIGURE 6.2 Pattern recognition of the human mind

happiness to the face (Figure 6.2C), and by adding two triangles one can convert the happy person into a happy cat, or at least some manner of animal (Figure 6.2D). Although the precise emotion and type of animal that are perceived may differ slightly by cultural context, the general process of pattern recognition holds for essentially all healthy humans who can see. Of course in reality, these shapes remain lines on a two-dimensional surface (they don't become an actual face or a cat in physical form); it is our perception that assigns the identity of a face or a cat to the different shapes. Like much of human perception, this is a "constructed reality" that we impose on the world around us. This reality often lines up with the actuality of the world, but it need not do so in every instance.[4]

The extent to which we will imbue two-dimensional abstractions with cognitive properties is substantial. I'm a big fan of Pixar animated films, and I appreciate how the movies have humor that appeal to children and at the same time make meaningful comic and

[4] Dr. Steven Novella has recorded a series of lectures on this topic that are highly informative and illustrate the case in clear terms. Several of the examples in this chapter are referenced and described by Dr. Novella in these lectures with a great clarity of presentation. I highly recommend these to the interested reader: "Your Deceptive Mind: A Scientific Guide to Critical Thinking Skills." The Great Courses.

artistic statements for an adult audience. Pixar also has the tendency to explore adult and existential issues, such as death and lost childhood. Recently, I was reduced to tears when I watched *Inside Out* with my daughter. Previously, *Toy Story 3* sent me into an existentialist tailspin (seriously, let's all just hold hands and face together the inevitability of death and destruction, so at least we aren't alone when the end comes?). What is interesting about both movies is that I assigned agency and personality to two-dimensional combinations of lines and dots – to completely abstract characters and not representations of real people. Yet, the degree to which my mind assigned human status to these images allowed me to experience my emotions as I would do when faced with real-life tragedy.

In some meaningful ways, one may draw parallels between the perception of patterns based on the human senses and the practice of retroducing hypotheses to explain natural phenomena, as was described in Chapter 2. When faced with a body of data, humans enter a creative process by which they "guess" at causes that would predict the already observed data. The perception of patterns in sensory input is not conceptually dissimilar, as the brain is guessing that a face already exists and is the cause of certain shapes. As mentioned earlier, there will be multiple hypotheses that each lead to the prediction of the same observed data. Interestingly, such is very much the case when the brain is assigning a pattern to a sensory stimulus. A classic example of this is the Rubin Vase, an image of a porcelain vase that one might put flowers in. However, when staring at the vase, the brain may perceive two different patterns (at least). The first is a vase. The other is two faces in profile looking at each other (Figure 6.3). Both perceptions are valid, and both are derived from the data at hand. Most people can see the alternate image when told how to look for it, even if they didn't initially notice it. It is very difficult for your brain to construct both images simultaneously (the actual content of the picture), and there are also many other abstract entities contained in the picture that we don't construct. The picture is what the picture is – it is each of these things (and many others that

FIGURE 6.3 Alternate perceptions of the same object

we don't recognize), but our brain tends to construct one pattern and specifically recognize that.[5]

I have given you a visual example, but the same principle works with the other senses. If you hear a voice on the phone that sounds like someone you know, you may assign an identity to that person's voice even though it may wind up being someone else. Similarly, if one hides an object in a box and you allow people to touch only a small part of it, their mind will construct an image of what the rest of the object is – at least what they imagine it to be.

[5] Visual images like the Rubin vase are often used to illustrate how different people can see the same world quite differently, an example of the notion of different paradigms that emerged with postmodernism. In the context of science, Thomas Kuhn's *The Structure of Scientific Revolutions* is best known as the basis of this movement. In the extreme form of this view, people who see the world as one paradigm are incapable of perceiving the other paradigm(s) and are also incapable of communicating with those in other paradigms (i.e., incommensurability). However, in the current example, the Rubin vase is used to show the "constructed" nature of experience. Most people can see either view when it is pointed out to them, but we tend to first perceive one or the other and tend to see either one or the other. The question "Is it faces or a vase?" is unanswerable, as it is simultaneously both and neither; a question that can be answered is "What do you perceive?" because our "reality" is a constructed reality based on our perceptions.

These examples have focused on a single sensory modality; however, we often experience entities through a combination of our senses, which typically work in an integrated fashion to allow our brains to construct our perceptions of the environment. However, curious things can occur when our senses are in conflict with each other. In 1976, McGurk and MacDonald published a paper in the journal *Nature* describing an experiment in which people are shown a film of a person talking, but the movement of the lips is inconsistent with the sounds being made (called the McGurk effect).[6] When this occurs, the sound that is heard is not the sound being made; rather, the sound that is heard is consistent with what the brain expects to hear based on the motions of the speaker's mouth. In a similar fashion, it has been reported that the addition of food coloring to give white wine the appearance of red wine causes some wine drinkers to describe the fragrances typically associate with red wines.[7]

The constructs that our brains construe from our sensory input are also highly influenced by our expectations of what we might experience. This reflects in many ways the suggestible nature of human perceptions. For example, when camping, if you're under the impression that you're sitting in a bug infested forest, you may start to feel bugs crawling up your legs or arms when no bugs are present. This is an example of the brain constructing perceptions from normal, small sensory sensations that are always being delivered to the brain. A famous example of this occurred in 1978 in Rotterdam in the Netherlands. A red panda was found to be missing from the Rotterdam zoo. The local newspaper published the story, hoping that people would keep an eye out for the little red panda and help zookeepers recover it. Over 100 reports of red panda sightings came in from

---

[6] McGurk H, MacDonald J. 1976. "Hearing Lips and Seeing Voices." *Nature* **264**(5588): 746–8.

[7] Morrot G, Brochet F, Dubourdieu D. 2001. "The Color of Odors." *Brain Language* **79** (2): 309–20.

around the city.[8] However, it was subsequently discovered that the panda had not actually escaped the zoo and was therefore not on the loose. It's unclear what the people were seeing (in their minds, they were clearly perceiving a red panda), but it was almost certainly not a red panda, which is indigenous to the Himalaya Mountains in China and Nepal and not northern Europe.

A second highly informative example occurred at the beginning of the Second World War. On December 7th, 1941, the Japanese fleet attacked and destroyed much of the American naval fleet at Pearl Harbor. Then, on February 3, 1942, a Japanese submarine attacked an oil depot at Ellwood, California. Americans on the west coast were on high alert and hypersensitive to an impending Japanese attack. On the evening of February 24, an air raid was perceived to be taking place over Los Angeles, alarms were sounded, and anti-aircraft artillery soldiers took their positions.[9] During that night and into the morning of the 25th, more than 1,400 artillery shells were fired at what the gunners believed to be Japanese planes. However, when the dust had settled, it became clear that no air raid had taken place and that there had not been any Japanese planes attacking Los Angeles.[10] It is unclear exactly what the gunners were firing at, for whatever they saw was not a force of attacking Japanese planes or any planes, for that matter. It could have been a chain reaction – when one gun fired, others began firing. Whatever it was, the brains of the gunners constructed a perception of Japanese fighter planes real enough to induce them into firing their guns at "objects" in the air.[11]

[8] Wynn C. 2007. "Seen Any Red Pandas Lately?" *Journal of College Science Teaching* **36**(5), 10–1.
[9] Wikipedia. n.d. "Battle of Los Angeles." https://en.wikipedia.org/wiki/Battle_of_Los_Angeles
[10] *Los Angeles Times*. Feb. 27, 1942. "Knox Assailed on 'False Alarm': West Coast Legislators Stirred by Conflicting Air-Raid Statements." p. 1.
[11] Dr. Steven Novella. "Your Deceptive Mind: A Scientific Guide to Critical Thinking Skills." The Great Courses.

PROBLEMS IN PATTERN RECOGNITION OF SENSORY
INPUT: PERCEIVING THINGS THAT ARE
NOT THERE

Humans are very good at constructing the perception of objects of a certain class, such as with faces and cars, as illustrated in the previous section. This ability extends to abstract art, such as an image that may depict a cat in a highly distorted or grotesquely deformed way, yet viewers still identify the image as a cat. This is literally a construct that the mind forms based on the data being received. Michael Shermer coined the general phrase "patternicity," which he defined as "the tendency to find meaningful patterns in both meaningful and meaningless noise."[12]

Noise is basically patternless data. Shermer points out that humans not only find patterns when patterns are there, but they also find patterns when patterns are not there. The downside is that our tendency to find a pattern is equally consistent with a pattern being there and with it not being there. Some people will hear voices in random radio static, but that is the brain assigning identity to something it can't quite recognize. Psychologists have named this general occurrence "apophenia," with "pareidolia" representing the specific act of finding meaningful patterns in random stimuli. A number of common examples of this exist, including seeing human faces on the moon or Mars, or perceiving a face in the bark of a tree. The act of finding recognizable forms in clouds is another example of pareidolia; the observer is cognitively aware that there really isn't a face or a bunny rabbit in the clouds, but the brain is continuously trying to fit the data it receives to known patterns, and allowing it to do so with clouds can be a fun practice. Pareidolia has given rise to numerous sightings of images of Jesus or the Virgin Mary in all manner of places, from discolorations on windows, to stains on grills in fast food

---

[12] Shermer M. 2000. *How We Believe: Science, Skepticism, and the Search for God.* London, UK: Henry W.H. Freeman and Company.

restaurants, to mold in shower stall corners.[13] In the 1960s, people thought Paul McCartney was dead because of words they perceived when they played Beatles' records in reverse. In such cases, the mind constructs patterns from random noise.

Of course, the vast majority of random noise is simply that; however, given enough time and by chance alone, recognizable patterns will form from it and human perception will latch onto it as something special. In fact, there is a practice known as electronic voice phenomena (EVP) – adherents listen to white noise static until they hear words emerge, which they believe to represent supernatural voices communicating with them. A well-publicized example of this was a toy baby doll recently sold on the American market (Fisher-Price's "Little Mommy Real Loving Baby Cuddle and Coo"). It was pro-grammed to make cooing noises that mimicked the sounds of a real human baby. However, pareidolia kicked in, and some people heard the baby say "Islam is the light." The resulting public outcry caused the toy to be pulled from the shelves.[14,15] In all fairness, I will point out that we cannot unequivocally rule out the presence of supernatural voices in static, nor can we say with certainty that the toy manufacturer didn't intentionally hide this message in the baby doll's voice. However, for the purposes of the current discussion, it is a fair statement that, given the body of psychological research on pareidolia, humans will find patterns where they don't exist and will do so frequently.

INTRINSIC DEFECTS IN HUMAN PERCEPTION:
MISSING THINGS THAT ARE THERE

In addition to finding patterns in random data that aren't actually there, humans also have a great ability to not notice what is right in

[13] Shermer. *How We Believe*

[14] Fox News. Oct. 9, 2008. "Parents Outraged Over Baby Doll They Say Mumbles Pro-Islam Message." www.foxnews.com/story/2008/10/09/parents-outraged-over-baby-doll-say-mumbles-pro-islam-message.html

[15] Dr. Steven Novella. "Your Deceptive Mind: A Scientific Guide to Critical Thinking Skills." The Great Courses.

front of them. Some people have an uncanny ability to recall details of their present experience; however, in general, humans miss a great deal of these details in any given situation. After passing through a room or witnessing a scene of some sort, it is not uncommon for people to have little knowledge or memory of color, patterns, shapes, etc. This failure extends not only to new or unfamiliar places; indeed, most people are hard pressed to describe the fine details of their bedrooms or living rooms (spaces where they spend large amounts of time on a daily basis). Of course, familiarity with an environment will result in having greater knowledge of it, compared with an environment only encountered once; nevertheless, most people only remember superficial details of familiar places. An image of the visual world is created on the retina and is transmitted to the cortex, but is not necessarily integrated into conscious observation.

Cognitive experimentation has demonstrated our weaknesses in detailed observation to a level that approaches embarrassment. A famous and much referenced experiment by Christopher Chabris and Daniel Simons involved a movie in which players were passing multiple basketballs back and forth.[16] The players were wearing either black or white uniforms, and observers were instructed to count the number of times that a ball was passed back and forth by players wearing black. During the film, a person in a gorilla suit walks across the screen, stops in the center, thumps its chest, and then walks off the screen. Yet only 54% of viewers noticed this! This effect has been termed "inattentional blindness."[17] You can watch the video yourself,[18] although knowing in advance what you are looking for will prevent inattentional blindness.

---

[16] Simons DJ, Chabris CF. 1999. "Gorillas in our Midst: Sustained Inattentional Blindness for Dynamic Events." *Perception* 28(9): 1059–74.

[17] Chabris C, Simons D. 2010. *The Invisible Gorilla, How our Intuitions Deceive Us.* New York: Crown Publishers.

[18] Chabris C, Simons D. 1999. "The Original Selective Attention Task." www.theinvisiblegorilla.com/videos.html

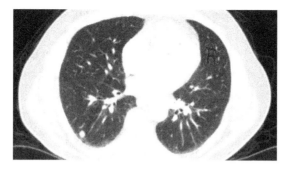

FIGURE 6.4 Inattentional blindness at work

Inattentional blindness is not just the result of seeing something for the first time (i.e., a basketball game with a guest appearance by a gorilla). The defect also extends to people who have spent countless hours observing the same type of scene over and over. For example, a study was conducted with radiologists who analyze CT scans in diagnosing lung cancer.[19] The radiologists were shown five CT scans with cancer nodules in them and asked to evaluate them. A small picture of a gorilla was placed on the upper right quadrant of the fifth scan (Figure 6.4). After the final scan was evaluated, the researchers asked the radiologists the following series of questions:

Did that last trial seem any different?

Did you notice anything unusual on the final trial?

Did you see a gorilla on the final trial?

Of the 24 participating radiologists, only 4 (17%) noticed the gorilla. The researchers had tracked the radiologists' eye movements, so they could reject the notion that the gorilla image simply hadn't entered the visual field(s) of the participants – it had. But despite the visual sensory input, the majority of participants didn't perceive the gorilla; they were looking for patterns that indicated cancer, and any

---

[19] Drew T, Vö ML, Wolfe JM. 2013. "The Invisible Gorilla Strikes Again: Sustained Inattentional Blindness in Expert Observers." *Psychological Science* **24**(9): 1848–53.

other details of what was right in front of them were filtered out somewhere between the visual input and their cognition.

The defect of not noticing details also extends to personal interactions with three-dimensional objects. One striking example is an experiment in which people meet another person whom they have not previously encountered (e.g., a salesperson behind a counter). When the subject is looking away, the salesperson ducks behind the counter and is replaced by a completely different person. The new person may be dressed similarly and have general features in common with the first person, but the two people are different individuals. Nevertheless, 75% of people in this experience do not notice the switch.[20] This is called change blindness, and it is another example of how much we misperceive in the world. This problem is not limited to vision, as 40% of people don't detect a change from one speaker's voice to another, halfway through a word list that is being read out loud,[21] A tactile form of change blindness has been described as well.[22,23]

A second problem, distinct from overlooking input, is the creation of incorrect input. Although the human mind does not collect sufficient detail to generate a seamless series of events, human perception of a particular event typically takes the form of an uninterrupted narrative. This occurs because the brain fills in details where actually observed details are lacking. Because the missing spaces are flanked on both ends with actual perceptions, the brain typically fills in the data correctly. This will happen when the brain fills in details during events that are happening too quickly for the brain to process.

---

[20] Simons DJ, Rensink RA. 2005. "Change Blindness: Past, Present, and Future." *Trends in Cognitive Science* 9(1): 16–20.

[21] Vitevitch MS. 2003. "Change Deafness: The Inability to Detect Change between Two Voices." *Journal of Experimental Psychology and Human Perception Performance* 29(2): 333–42.

[22] Gallace A, Tan HZ, Spence C. 2006. "The Failure to Detect Tactile Change: A Tactile Analogue of Visual Change Blindness." *Psychonomic Bulletin & Review* 13(2): 300–3.

[23] It is difficult, if not impossible, to illustrate the full effect of such effects in written text; however, the interested reader can experience stunning examples of this in the popular television show "Brain Games" that shows videos and examples.

Baseball players regularly make contact with fastballs travelling at more than 90 miles per hour. In actuality, baseball players can only perceive parts of the ball's flight, but they remember seeing a continuous pitch. They are able to swing the bat to hit the ball even though they can't process visual images quickly enough to see the bat striking the ball. Their brains fill in these missing details.

The process of filling in details is not restricted to creating a continuum for moving objects, but rather extends to all kinds of observation. It is well documented that many eyewitnesses will provide different and contradictory descriptions of the same event. They received the same visual information in terms of light hitting their retinas, but their brains generated very different images and they ultimately "saw" different things ("saw" being defined as the cognitive function of perceiving or constructing an event in the human brain). What is of particular importance is that many witnesses will believe, absolutely and unequivocally, that his or her version of events is the correct version.

Our ability to fill in details, or to overlook gaps and defects, can be dramatically illustrated by the ability of most individuals to read the following passage:

> Aoccdrnig to a rscheearch at Cmabrigde Uinervtisy, it deosn't mttaer in waht oredr the ltteers in a wrod are, the olny iprmoetnt tihng is taht the frist and lsat ltteer be at the rghit pclae. The rset can be a toatl mses and you can sitll raed it wouthit porbelm. Tihs is bcuseae the huamn mnid deos not raed ervey lteter by istlef, but the wrod as a wlohe.[24]

LACK OF FIDELITY IN HUMAN MEMORY

The ability to remember is essentially the only thing that makes us cognitively four-dimensional beings. We are not four-dimensional

---

[24] The exact origins of this text appear to be unclear. It was first circulated on the Internet in September of 2003, and it has been analyzed by a number of groups, including researchers at the Cognition and Brain Sciences Unit at Cambridge University, www.mrc-cbu.cam.ac.uk/people/matt.davis/cmabridge/

creatures in our perception. We perceive three dimensions of space and an instant of time at any given moment, but we don't experience time-depth. We can only experience an instant; we don't experience an entire minute simultaneously. In other words, without memory there can be little to no conception of time. Likewise, in the absence of memory, there can be no induction: One can't predict the future based on the past if one has no knowledge of the past, and one can have no knowledge of the past without memory. Even if the past is recorded, you must have sufficient short-term memory to hold the recorded material in your mind long enough to think about it. Finally, the synthesis of any higher order of understanding necessitates the simultaneous association of multiple individual factors; if one can't remember components of a system, then one cannot put them together. Memory is essential to our cognition and our ability to think anything other than an immediate reflex response. Without memory we couldn't think past the sliver of time that is now, and so all we could do without memory is respond to stimuli reflexively; there could be no learning or continuity.

Each of us has a life narrative based on our recollections of our experiences, but the accuracy of our narrative is a function of the fidelity of our memories. What would be the implication if our memories were actually highly flawed? What would it mean if we forgot the majority of the things we experienced, and our minds modified the memories we retained by changing them each time we remembered them? What if we took stories we heard from other people and imagined they happened to us, or if our minds confabulated things that simply never occurred? What would it mean if someone was told that what they remember is wrong, and they just changed what they remembered to fit the new information instead of questioning what they were being told? What would it mean if people assigned truth to remembered statements because they sounded familiar, without attention to whether the statements were actually true?

The humbling and somewhat frightening situation is that while human memory can have very high fidelity, it also tends to make each of the errors just described. Worst of all, the confidence we have in our memories is not directly linked to the likelihood they happened or to the accuracy with which we remember them. We tend to be highly confident about our memories, even when they are unrelated to what actually occurred. In December of 2015, during his run for the presidency, Donald Trump made the claim that groups of people in the streets of New Jersey were celebrating the collapse of the World Trade Center on September 11, 2001. Members of the media scrutinized this claim with great detail, found no evidence that this had ever occurred, and determined that this had been a false rumor. However, Trump refused to back down on his claim, citing as evidence that he "saw it on television somewhere," that hundreds of people who witnessed the event had called or tweeted him, and that the *Washington Post* had commented on the celebrations in one of its articles at the time. However, a careful examination of the record shows no indication or reporting of such an occurrence; at best there are some references to rumors to this effect. The existence of rumors has not been disputed; however, there is no indication that the rumors have any basis in an actual occurrence. Many have called this belief by Donald Trump a lie, but in all likelihood, Donald Trump actually "remembers" people celebrating the destruction of the World Trade Center and is very confident about his memory, despite a strong amount of evidence that such an event never occurred. The logical impossibility of demonstrating the absence of something is noted here as, "the absence of evidence is not the evidence of absence." However, unless there is a broad conspiracy, it is a reasonable prediction that there would be at least some reporting in the media if such an event had actually occurred. Whatever we think of Trump and his memory, "misremembering" is very common. We each likely remember something that didn't occur, probably many things that didn't occur, but upon which we would each bet our lives.[25]

---

[25] Since this chapter was first written, President Trump has made many more claims of "facts" for which there is no evidence at all (or evidence to the contrary). Sadly,

In addition to incorrect details of a memory, we are highly susceptible to altering our memories as a result of experiences that occur after the event being remembered. A classic study was performed by Loftus, Miller, and Burns in which people were all shown the same two pictures and then asked a series of questions. Depending upon which questions people were asked, they remembered the details of the pictures differently. This has been replicated multiple times, in what has been called the "misinformation effect." When people are instructed that their memories are incorrect, it represents a sort of breakdown of coherence. In many cases people reestablish coherence not by denying what they have been told is true, but rather by modifying their memories to conform with what they were told. Since memories amount to stored observations, people alter their observations to conform to authority instead of questioning authority based on what they observed.[26] Once a memory is altered, then the next time one thinks of it, one will remember the last time it was remembered, and in this way, it becomes a "real" memory even though it never occurred.

Regrettably, the issue goes much deeper than just modifying existing memories. Indeed, entirely fabricated, detailed memories can be implanted in people by having family members tell them of events that allegedly occurred to them, when in fact the events never happened. These are called "rich false memories," and examples include the following: remembering being lost in a shopping mall at 6 years of age and then being found, being hospitalized with a particular illness, suffering an attack by an animal, and a near-drowning

the list just keeps growing. However, the growing of the list does not distinguish between a president who is purposefully lying and one who is falling into a cognitive trap of memory. If it is the latter, the implications of the present chapter grow ever more profound, especially if our elected officials don't acknowledge problems with human memory.

[26] This is reviewed in the article, Loftus EF. 2005. "Planting Misinformation in the Human Mind: A 30-Year Investigation of the Malleability of Memory." *Learning and Memory* **12**: 361–6.

experience.[27] It has also been shown that certain tactics of police interrogation can cause suspects to create rich false memories of crimes that they did not commit.[28] In some cases, highly detailed and compelling memories are created. Moreover, in some cases it can be shown that rich false memories cannot have occurred, and the only viable source of the memories is that they have been created de novo – that they are not based on forgotten events. Thus, concerns about memory extend not only to the accuracy of remembered details, but to complete fabrication as well.

People also have a tendency to mistake the source of the information they are remembering as well as the content. Because of these errors, memories are highly susceptible to manipulation by suggestion. A number of studies have demonstrated that people who are asked to imagine an event that never occurred recall "actual" memories of the event more frequently than those who were not asked to imagine it.[29,30] Finally, people have the tendency to assign truth to claims that seem familiar, regardless of whether they were told the claims were true or false when they encountered them. Strikingly, even when people are exposed to repeated claims that a particular thing is false, they have the tendency to later remember it as being true just because it sounds familiar.[31]

Thus, even if humans perceive the world properly, even if they see only what is really there (which is clearly not the case), it appears that we nevertheless distort accurate perceptions over time because of problems in how our memories function.

[27] Loftus EF, Bernstein DM. 2005. "Rich False Memories: The Royal Road to Success." In Healy AF (Ed.). *Decade of Behavior: Experimental Cognitive Psychology and its Applications.* pp. 101–13. Washington, DC: American Psychological Association.

[28] Shaw J, Porter S. 2015. "Constructing Rich False Memories of Committing Crime." *Psychological Science* **26**(3): 291–301.

[29] Garry M, Manning CG, Loftus EF, Sherman SJ. 1996. "Imagination Inflation: Imagining a Childhood Event Inflates Confidence That It Occurred." *Psychonomic Bulletin & Review* **3**: 208–14.

[30] Goff LM, Roediger III HL. 1998. "Imagination Inflation for Action Events: Repeated Imaginings Lead to Illusory Recollections." *Memory & Cognition* **26**: 20–33.

[31] Skurnik I, Yoon C, Park DC, Schwarz N. 2005. "How Warnings about False Claims Become Recommendations." *Journal of Consumer Research* **31**: 713–24.

MISPERCEPTIONS BASED ON HUMAN SENSES

Human senses are highly adapted to detect complicated input in the context of navigating our environment. However, this adaptation can lead to striking misinterpretations in certain situations, which constitute perceptual illusions. Illusions are things that are misperceived, and these misperceptions tend to remain even after they are shown to be misperceptions. An excellent example of this can be demonstrated by a now popular picture that shows colors in the context of different shades. In Figure 6.5 there are two separate panels, one on top of the other; the top panel has a uniform background and the bottom panel is shown over a complex background texture.[32] To most people's eyes the upper panel is a darker gray than the lower panel. However, the difference in color between the two panels is an illusion.

A.   B.   C.

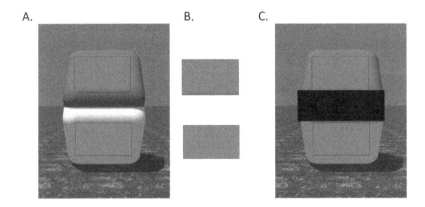

FIGURE 6.5 Visual illusions that persist even if the viewer knows they are there. Republished with permission of Society for Neuroscience. Copyright © 1999 Society for Neuroscience; permission conveyed through Copyright Clearance Center, Inc.

---

[32] The original picture is in color and has been changed to black and white here, but the effect persists. The original can be found in Purves D, Shimpi A, Beau Lotto R. 1999. "An Empirical Explanation of the Cornsweet Effect." *Journal of Neuroscience* **19**(19): 8542–51.

In this particular case, the two panels are an identical color – the same shade of gray. If one moves the portion of the panels shown in black boxes out of context of the shadowing (i.e., moved to the right, Figure 6.5B), the colors can now be seen to be the same. Similarly, if one covers the center shadowing with a black box, the colors now appear the same (Figure 6.5C). One can cover the center shading on the first figure to ensure no trickery. The human mind has learned that colors in relative shade or in shadow tend to appear darker than they really are, whereas colors in bright light tend to appear lighter than they actually are; the human mind compensates accordingly. Yet, despite knowing this, the original image still has the appearance of panels of different colors. This is a simple example – just the tip of the iceberg – of how persistently flawed our human senses can be.[33]

OBSERVATIONS ARE "THEORY LADEN" AND HIGHLY INFLUENCED BY BACKGROUND BELIEFS

The historian and philosopher Thomas Kuhn turned the study of science entirely on its head with the publication in 1962 of his landmark work, *The Structure of Scientific Revolutions.* One of the many ideas put forth by Kuhn is that what we observe is directly affected by our background knowledge and beliefs. In other words, what one observes is a function not only of sensory input but of the theories one has in mind when making the observation; observations are "theory-laden." In many ways this seems incompatible with a system where one evaluates theories and ideas, and accepts or rejects them based on what one observes. Indeed, if observations are a result of what one believes prior to the observation, or are at least strongly influenced by such beliefs, then how can we ever use an observation to assess a belief?

In some ways, Kuhn's idea is not that revolutionary; most of us have probably observed this phenomenon firsthand.[34] One can receive

---

[33] One can access a great number of visual illusions, both static and in motion. One excellent source is the popular television show "Brain Games."

[34] Kuhn certainly called this problem out in ways no one else had. It should be noted that a previous publication by Ludwick Flick called "Genesis of a Scientific Fact"

an identical, word-for-word email from two different people; the email will seem hostile if it is received from a person with whom you have had historical animus, while it will seem warm if it's from a long-time friend. You experience the same email differently based on your background belief of what the author's tone is likely to be. Your preexisting experience and beliefs alter what you observe and how you observe it.

Kuhn developed this notion across the history of science in a way that was striking. He argued that scientists are trained into certain paradigms, and thereafter, the paradigm into which they are indoctrinated alters how they perceive the world. Consider two separate scientists, one of whom was raised to believe in an Earth-centered solar system and the other in a Sun-centered solar system (two different paradigms). Each of them sees the sunrise[35] one morning. This is not an illusion, they both receive the same general images with their eyes, but based on their world view, they perceive different things. One of them sees the Sun going around the Earth, the other sees the effects of the Earth rotating.

A classic example is found in the history of my own scientific field, that of blood transfusion. One of the first recorded transfusions happened in France in 1667. In this case, a 34-year-old patient was brought to his physician to be treated for "running naked through the streets of Paris; a 'phrensy' brought about by the mental anguish of a bad love affair." Given the theory of human diseases at the time, the man's condition was determined to result from an imbalance of his humors and having too much "vice" in his blood. Wanting to flush this vice out of his system, his physicians thought a transfusion might do the trick. However, they believed that all humans were born in "original sin" due to that regrettable incident with the apple in the

---

partially identified this issue. Kuhn acknowledges its influence on him in the introduction to *The Structure of a Scientific Revolutions*.

[35] The common English word "sunrise" seems to imply that the Sun goes around the Earth – otherwise we would call it "horizonsink."

garden.[36] Thus, one could not transfuse the patient with human blood, as this would just be adding vice on top of vice. Rather, it was determined that a calf's blood should be used, as animals do not have original sin. After the first treatment the patient seemed to have improved. A second transfusion was then given, resulting in what is now considered a classic description of a type of biological reaction to transfusion.

> As soon as the blood began to enter into his veins, he felt the like heat along his arm and under his arm-pits which he had felt before. His pulse rose presently, and soon after we observed a plentiful sweat over all his face. His pulse varied extremely at this instant, and he complain'd of great pain in his Kidneys, and that he was not well in his stomach, and that he was ready to choak unless they gave him his liberty... He was soon made to lie down ... and slept all that night without awakening till next morning ... When he awakened ... he made a great glas full of Urine, of a colour as black, as if it had been mixed with the soot of Chimneys.[37]

In modern medical understanding, what this poor individual suffered is called a "hemolytic transfusion reaction." Basically, his immune system had responded to the initial transfusion by making antibodies against calf's blood, such that upon the second transfusion the blood cells were rapidly destroyed by the immune system. This is now well known to cause all the signs and symptoms described – the black urine is the result of the released contents of destroyed red blood cells passing through the kidneys. This can cause horrible kidney damage, and toxicity to multiple organ systems can ensue.

---

[36] There has been much debate that if a fruit were to be found in the garden of Eden, it would more likely be a fig.

[37] Denis J. 1668. "An Extract of a Letter: Written by J. Denis, doctor of physick, and professor of philosophy and the mathematicks at Paris, touching a late cure of an inveterate phrensy by the transfusion of bloud." *Philosophical Transactions of the Royal Society B: Biological Sciences* **4**: 710. https://royalsocietypublishing.org/doi/pdf/10.1098/rstl.1666.0065

Instead of being horrified by the patient's symptoms and the tissue damage that was spilling out through his failing kidneys, the physicians were delighted at this result. They interpreted the black urine to be clear evidence of vices being flushed from the patient's body. What they observed was what they had expected to see – the result of the theory they had in mind at the time the reaction occurred. The physicians were so encouraged by the vices being flushed from his body that they later gave a third transfusion, from which the patient died.

In the more extreme form of the argument, Kuhnians may contend that humans literally cannot observe data that contradicts their theories, because the theory determines how they see things. Moreover, scientists who operate from within different paradigms literally cannot communicate with each other, both because they may use the same words for different meanings (and not even know they are doing so) but also because they cannot observe the same natural world.[38]

For the purposes of the current discussion, it is most important to understand that observations are theory-laden, and as such, how we observe the natural world is influenced by what we already believe – a further blow to the ivory tower idea that science is a series of concepts that are clearly sorted out by purely objective evidence. If evidence is theory-laden, it cannot be entirely objective. However, this should not be confused with evidence and observation having no objectivity at all. Rather, the effect of theory on observation should be another source of potential error that, now that science is aware of it, we should seek methods and approaches to mitigate its effects.

---

[38] The implications of these concepts are profound and go very deep, affecting issues of the nature of scientific communication, ideas, and concepts; the nature of progress; and development of scientific ideas. The interested reader is encouraged to explore the rich literature on this area of scientific postmodernism and the debates to which it has given rise, which have dramatically been designated the "Science Wars."

A GENERAL PROBLEM WITH ALL EXPERIENCE-BASED KNOWLEDGE FROM OTHERS AND OURSELVES

Information of which one has no direct experience is fundamentally problematic. For example, one reads of an event in the newspaper and thus gains evidence of the event. However, one must now rely on the accuracy of the news story, that the journalist correctly interpreted the event described, that the witnesses themselves perceived the event accurately, that there was no miscommunication between the witnesses and the journalist (including any intermediaries), and that the article wasn't edited or altered in a way that inadvertently changed the meaning. In addition, one must consider whether the journalist and/or newspaper is purposefully changing the description of what happened to fit an underlying agenda, be it political preference, an incentive to sensationalize news to sell more newspapers, or having other ulterior motives.[39]

While information may be purposefully distorted or altered, the regrettable reality of the situation is that information will be distorted even in the complete absence of any intent to do so. The popular children's game called "telephone" beautifully illustrates the fundamental problem of receiving information from other people. The children sit in a circle and pass a whispered message to one another. By the time the message reaches the last child the words and meaning have typically changed dramatically, often bearing little resemblance to the initial message (even with no intent to distort). This reveals why one should be at least somewhat skeptical (at some level) of any information that one receives from another person or source, which constitutes the vast majority of information we receive. Each of us, everyday, is involved in one massive game of telephone.

In some ways, one can extend this problem of information from other sources to include our previous selves as a source. As such, this

---

[39] This issue has been at the heart of recent arguments in the American political system about what the facts of a particular situation actually are and what the term "fake news" means.

relates to the earlier discussions of memory and all of its flaws. Quine pointed out that, "[A]n observation sentence ceases to be an observation sentence, after all, when we change the tense of its verb. Reports of past observations involve inference..."[40] In other words, other than what you are perceiving right now, all information you have ever gathered in the past is subject to problems of memory distortion, and thus the reliability of information decreases the moment it is not currently experienced. Your current observations may be flawed by errors in perception, and now there is the additional problem of remembering correctly, which is a real concern because the data indicate that humans misremember horribly even if they perceive correctly in the moment.

Thus, even our personal experiences and not just what we learn from others, suffer from this general problem. The "knowledge problem" introduced in Chapter 1 extends not only to problems of inducing things not yet observed or encountered, but also to data we or others have previously observed and reasoning we or others have previously done. We may not correctly perceive that which we experience as it's happening, but even if we did, we also distort it further over time as our memories reconstruct what we initially observed and another round of distortion is superimposed when we communicate the information to other people. When an event or fact is recorded, the closer the recording is in time to the event, the less likely such problems are to manifest. Conversely, the further away a record is in time from an event, and the more it has been handed down through oral tradition or storytelling before being recorded, the more suspect it should be and the more likely it is to contain misinformation and/or complete fabrication.

A LACK OF RECOGNITION OF THE PROBLEMS IN HUMAN OBSERVATION

In light of recent cognitive psychological experiments that have confirmed the unreliability of eyewitness testimony, an effort has been

---

[40] Quine WV, Ullian JS. 1978. *The Web of Belief*. New York: McGraw-Hill Education.

made to afford such testimony a lower evidentiary weight in the American legal system. In *Convicting the Innocent: Where Criminal Prosecutions Go Wrong*, author Brandon Garrett examines the cases of 250 convicted criminals who were ultimately exonerated by DNA evidence. Of the 250, 190 were convicted predominantly or exclusively by eyewitness testimony. One could take a cynical view and argue that many of these witnesses intentionally lied, but given what we know about problems with human perception, that need not be and almost certainly is not the case.

In *Commonwealth* v. *Walker* in 2008, a Pennsylvania case, a person was convicted of committing a crime at night on the basis of eyewitness testimony. As a third party to the trial, the American Psychological Association (APA) filed an amicus brief (friend of the court) showing evidence for the unreliability of eyewitness testimony. However, the judge would not allow the jury to hear expert testimony on this topic. The jury was kept in a state of ignorance about the unreliability of the evidence upon which they were deciding a person's fate.

A related and subsequent case was heard by the United States Supreme Court (*Perry* v. *New Hampshire*, No. 10–8974). In this case, an eyewitness testified that she observed Barion Perry breaking into a car and stealing its contents in a parking lot in Nashua, New Hampshire. The witness described seeing "a tall black man" with no other specific details. Moreover, she was unable to pick Perry out of a lineup. Nevertheless, Perry was convicted. The specific argument was that her identification was made under conditions that "suggested" Perry to her. His conviction is questionable precisely because of the strong body of evidence that human memory is highly susceptible to suggestion. However, the law provides an exception only when the police intentionally create a "suggestible" environment. In this case it was argued that a suggestible environment existed and thus undermined the credibility of the testimony, regardless of whether the police engineered it or not. In an 8 to 1 vote, the Supreme Court ruled against Mr. Perry. The majority decision by Associate

Justice Ruth Bader Ginsburg stated: "We do not doubt either the fallibility or the importance of eyewitness evidence generally...In our system of justice, however, the jury, not the judge, ordinarily determines the trustworthiness of evidence." As the sole dissenting voice, Associate Justice Sonia Sotomayor wrote: "[J]urors find eyewitness evidence unusually powerful and their ability to assess credibility is hindered by a witness's false confidence in the accuracy of his or her identification. That disability in no way depends on the intent behind the suggestive circumstances."

Thus, the aggregate effect of these two judicial opinions is that the justices are well aware that eyewitness testimony is unreliable, and they are equally aware that juries do not understand the unreliability of it. Nevertheless, they are unwilling to instruct juries on how unreliable such testimony is, and do not discard eyewitness testimony known to be generated under suggestive circumstances, unless such circumstances are intentionally generated by the police. The legal community has been slow to acknowledge this issue, although some recent progress has been made and will hopefully continue. As shall be described in the last section of this book, science continues to develop more advanced strategies for tackling this problem head-on and mitigating its effects. To be fair, this may simply be a function of the law being unable to revisit the actual crime itself as it happened, often eyewitness testimony is all they have. In contrast, scientists repeating experiments is a kind of revisiting a crime. As discussed in the previous chapter, reproducibility of experiments is essential to science and part of what distinguishes it. It is also possible that the implications of judicial acknowledgment would be to bring into question every conviction that has used eyewitness testimony of a certain type, and this would be an impossible political event for the courts, because so many cases involve precisely this kind of testimony. Nevertheless, given the credo of preferring to set the guilty free than convict the innocent, the slow adaptation of our legal system to the clear psychological evidence is regrettable. Along this theme, it will be argued later on that a progressive development of methods to remedy previously unperceived errors, as they become known, is a distinguishing feature of science.

FILTERING THE NATURAL WORLD THROUGH SCIENTIFIC INSTRUMENTATION

For many centuries, humans have made use of tools that translate and bring into our perception natural phenomena that we can't perceive with our senses. In some cases, this consists of simply amplifying signals that feed into our normal sensory inputs (e.g., telescopes and microscopes can bring into clear view that which is too far away or two small for our eyes to perceive on their own). Other instruments transpose signals that we cannot perceive into ones that we can observe. Some of these take the form of expanding the reach of our current senses, such as creating visible images based on the ultraviolet spectrum of light or converting sounds that are normally outside the range of what human ears can hear into audible signals. Alternatively, some instruments measure properties for which we have no sensory capacity at all and convert them into that which we can observe (e.g., sensors of electromagnetic fields, radiation detectors, particle detectors for subatomic entities).

The extent to which such instruments are actually reflecting reality has long been questioned. When Galileo utilized a telescope to see moons around Jupiter, people wondered whether the moons were an optical artifact created by the telescope itself. This general objection can be legitimately extended to all instrumentation. When physicists claim to have detected the Higgs boson, no human actually detected such a thing. Instead, a massive and complicated instrument delivered a signal that scientists interpreted as the effect of a boson existing. How are we to know if such a signal actually represents the natural world or if it is a machine-introduced entity? Additionally, even if it does reflect something in the natural world, how are we to know if our interpretation of what it reflects is correct?

The objections and concerns regarding instruments is a complicated subject, yet some progress has been made through the methodological assessment of instrumentation. As described previously, human sensory organs are themselves prone to error, as is the brain that interprets their input. Thus, the problem of instrumentation introducing artifacts is really only an extension of our concern about our own sensory organs. The combination of instrument error coupled with human sensory error just compounds the problem.

In this chapter we explored the mistaken assumption that humans can trust their perceptions of the natural world. Research in recent decades has shown this assumption to clearly be in error. Our senses certainly appear to have a strong link to whatever reality is out there, or else how could we navigate whatever world there is? Yet, our senses and our memories thereof also fail to accurately reflect reality, a situation of which we may be unaware. How science can mitigate such defects and how such mitigation helps to define scientific practice, will be described in the last section of the book.

For now, as long as we are defining the problems before us, let's set these problems aside and focus, in the next chapter, on an additional concern – errors in detecting associations among things. Even if induction was justifiable, our deductions and reasoning were always correct, if retroduction was not fallacious, if our observations were 100% correct all of the time, and our memories had perfect fidelity, errors of association would still exist.

Some References and Suggested Material. *In seeking background information for this part of the book, I was highly influenced by a series of lectures given by Dr. Steven Novella entitled "Your Deceptive Mind: A Scientific Guide to Critical Thinking Skills." While I have attempted to access source material and provide primary references, I was made aware of many of the examples in this chapter from Dr. Novella's lectures, which I recommend highly to the interested reader. Likewise, the popular television series "Brain Games" does an outstanding job of illustrating many of these principles, using video and visual tools not available in book form. The interested reader is encouraged to access these sources, and the primary research upon which they are based.*

7 Detection of Patterns and Associations, or How Human Perceptions and Reasoning Complicate Understanding of Real-World Information

Who wouldn't like to make some extra money? And if you have some money to invest, who wouldn't like to invest it wisely? But does any financial expert have a strategy that can consistently result in higher returns than the simple market average; i.e., an active investing strategy that can do better than just buying an index fund across a financial sector and then leaving it alone? Basically, is there anyone who can "beat the market"?

Of course, many experts and books claim the ability to do so, but the details may be scant and hard to figure out. To be fair, certain investors and hedge fund managers seem to have a real talent for making money. Time and again, year after year, in different environments and economies, they consistently outperform market averages. In other words, they beat the market and can take strategic advantage of a changing world. For the purposes of this discussion, we shall assume that they truly do so without using illegal insider information; although we know that this is not always the case. However, there are certainly many honest and scrupulous investors who consistently make money.

Haven't these investment experts achieved the goal of scientists, to develop a theory or understanding that gives the ability to predict? It's very hard to argue against this, given the data. Doesn't this look like the ideal of a system of scientific understanding? Those experts who can predict unanticipated economic developments have gotten it right time after time.[1]

There is, however, a cautionary tale here. Whether it's predicting the next bull market or the next major earthquake, humans are horribly bad at telling whether their predictions are the real result of understanding something about the system, or if the correct predictions are happening by chance alone. We are gaining an increasing understanding of the different mechanisms by which humans make very serious mistakes in this regard. Even when setting aside the problems detailed in the last chapters and even if humans make perfect observations of events and data, humans have an additional level of error in interpreting patterns and in determining when an association between two things actually exists. As we shall explore in this chapter, our baseline tendencies and common sense often betray us in this regard.

MISTAKEN LIKELIHOOD AS A SOURCE OF HUMAN ERROR

During my senior year of high school, a close friend and I frequented the horseraces at the track in Arlington Heights, Illinois.[2] One day a 30-to-1 long shot won its race (sadly, the name of the horse escapes me). After the race ended, two older men walked past us. One of the men had lost his money after betting on the favorite and was ranting at his companion: "A 30-to-1 long shot won the race. A 30-to-1 long shot! For Pete's sake, what are the odds of that happening?" Of course, the actual odds of a given horse winning a race are impossible to determine; the posted odds at the track merely reflect the sentiments of the betting public. Still, this story does betray an underlying

---

[1] There is some debate to what extent finances and economics are a science, and if so, to what extent and what type of science. However, this issue is not relevant to the current discussion – the reader can view economics as a science or not, with no effect on the points being made.

[2] As a matter of record, I was 18 at the time, the legal age for betting at a racetrack in Illinois.

misunderstanding of odds and probability, and how people are drawn to think about and examine unlikely events.

The occurrence of unlikely events catches our attention because they violate our expectations of how the world works – they challenge our HD coherence. By definition, very rare events do not happen often and thus they are not what we predict or expect. Accordingly, unlikely things cry out for and require explanation; they couldn't have just happened by chance, or could they?

From a positive point of view, extremely unlikely events are often perceived as "miracles" (e.g., the spontaneous remission of a cancer or the survival of a person who falls from an airplane without a parachute). Because it seems so unlikely (and perhaps impossible) to have occurred by chance, people feel compelled to seek an alternative explanation. Often a divine miracle is declared – the person was "meant to live" or it "wasn't his or her time." Perhaps in the grand plan of things this person has some special task or work to do, and cannot die before it's completed. It's common for the person who survived to feel as if he or she survived for some reason or purpose. This may change how the person lives his or her life, what he or she does with personal resources, and how he or she is treated by others. People who know of the incident may change their behavior, how they pray, or how they view the world. In extreme cases people may look to the person for wisdom, touch them to get a "blessing," or even worship the person in some way.

Negative outcomes are no less in need of explanation and maybe even more so. A person who suffers from an unlikely negative occurrence may consider him- or herself cursed. In secular terms, perhaps they are categorized as a fundamentally unlucky person. In supernatural terms, there may be unseen forces out to get them, perhaps spontaneously or as punishment for past actions. In more formal religions, the person may have lost the favor of his or her god and is being punished or tested in some way. We struggle to make sense of extremely unlikely things that do not make sense to us. However, in many cases, that which seems extremely unlikely is actually not so improbable and may even be highly likely to occur. In such cases, human misperception of probability results in seeking a cause to explain an unlikely event when there is no need or basis to do so. But why would something that was likely to occur appear to be highly improbable?

GENERAL EFFECTS OF EXTENSION NEGLECT

Psychologist Daniel Kahneman has coined the term "extension neg-lect" to capture a series of findings that demonstrate an underlying tendency of humans to evaluate data without consideration of the background information and context.[3] It is so unlikely that any given person will win the lottery that when someone actually does, people want to know where that person bought the ticket, what clothes they were wearing that day, what they had for breakfast, etc. This is due to the perception that the odds of any given person winning the lottery are so infinitesimal that winning requires an explanation as to what might have caused it. However, while it is true that the odds of any one particular person winning the lottery are low, so many people play the lottery that the odds of at least one person winning are pretty high. We tend to focus on seeking an explanation of why a particular person won, but an explanation would also be required if no one won the lottery for a long period of time – it would seem as though the system was rigged to avoid anyone winning. Focusing on those who won the lottery and ignoring all of those who played but didn't win is called "the lottery fallacy." The fact that someone won the lottery is not an unpredicted event, is not unusual, and does not require any kind of explanation – given the large number of people playing.

Extension neglect has very real effects on broader human behav-ior. Consider when someone sees an image of Jesus Christ or the Virgin Mary in the discoloration of glass, or in a burn pattern, or on

[3] Kahneman D. 2000. "Evaluation by Moments, Past and Future." In Kahneman D, Tversky A. (Eds.). *Choices, Values and Frames*. p. 708. New York: Cambridge University Press.

a restaurant grill, or in the clouds. Pilgrims flock to these sights, as they are interpreted as a sign of a divine presence manifesting itself in mortal affairs. It is much longed for evidence of the presence of a deity and that we are not alone, that our greater inclinations regarding divinity are correct. Indeed, it seems unlikely that "randomly" distributed matter should coalesce into the image of anything without some intervention. However, if one considers the number of cloud formations that occur in which Jesus is not seen, the number of distortion patterns in glass that have no discernable image, the number of grill patterns that are ignored, then the appearance of Jesus in these isolated cases seems much less significant. Indeed, if one examines enough random things, images of Jesus will appear by chance alone, as will images of the flying spaghetti monster and images of my pet ferret Waldo. Images of anything one is looking for will appear by chance alone, if enough things are examined, and an endless number of things are examined every day by approximately 7 billion humans existing and encountering their world. Yet people flock to the hits and ignore the multiplicity of misses without consideration for the background information.

There was a kind of a panic in the United States in 1985 when a rumor arose that participating in the fantasy role-playing game Dungeons and Dragons (D&D) was causing adolescents to commit suicide. Indeed, in that year there were reports of at least 22 teenagers who were highly engaged in this game and who did end up committing suicide. The media caught hold of this story and anxiety began to brew in American society. The television news show "60 Minutes," which had a viewership in the millions, did a full story on the issue. At first glance, 22 teen suicides certainly is a very scary number that appears to require some kind of investigation. However, further study revealed that, at the time, more than 3 million teenagers played D&D. The rate of suicide among teenagers in the United States in 1985 was 12 per 100,000; therefore, one would predict that by chance alone, of the 3 million teenagers playing the game, 360 should have killed themselves. Thus, upon further examination, it actually appeared that the game had a protective effect in preventing *more* teens from killing themselves. Despite this, some townships went so far as to enact laws that banned the game.[4] This example once again shows how the meaning of an observation changes profoundly based on background information – but humans often don't consider the background information.[5]

MORE EXAMPLES OF THE LOTTERY FALLACY IN EVERYDAY LIFE OTHER THAN THE LOTTERY

The national weather service estimates the chance of an average American being struck by lightning in a given year at approximately 1 in 500,000. This indicates that with a population of about 350 million, approximately 700 Americans will be struck by lightning annually. Having this happen to you would certainly be considered very bad luck. However, assuming an average lifespan, there is a lifetime chance of being struck by lightning of 1 in 6,250, which is a much more likely occurrence[6].

The more potent consideration is that the odds of a person being struck by lightning twice in his or her life are estimated at 1 in 9 million. This is a population average that exceeds the joint probability of two lightning strikes hitting any one individual at random, because the actual chance of being hit by lightning is affected by the part of the country you live in, whether it's rural or urban, and also one's behavior during thunderstorms. On the flip side, the odds are also decreased somewhat because a certain percentage of people die the first time lightning strikes them. Given the U.S. population of about

---

[4] *BBC News Magazine.* April 11, 2014. "The Great 1980s Dungeons & Dragons Panic." www.bbc.com/news/magazine-26328105

[5] This particular D&D example was used by John Allen Paulos in his excellent book entitled *Innumeracy: Mathematical Illiteracy and Its Consequences*, which shows how bad humans are at assessing probability. Paulos JA. 2001. *Innumeracy: Mathematical Illiteracy and Its Consequences.* New York: Hill and Wang, pp. 168–9.

[6] There are several different sources that estimate lifetime risk of being hit by lightning, and the precise estimates vary to some extent, but are all in the same general range.

350 million, this means that approximately 39 Americans alive today will actually be struck by lightning twice in their lifetimes.

When someone has been twice struck by lightning, it really catches our attention. Of course, we don't hear reports from 350 million people each day confirming that they still have not been struck by lightning twice. Thus, we focus on the hits without considering the base rate of misses, and we seek an explanation for what seems unlikely to happen by coincidence alone. Indeed, for the twice-struck person, simple bad luck of random things would likely no longer be the explanation. The common expression that "lightning seldom strikes twice" suggests there is a special cause affecting such a person. Rather, people with such an experience may look for alternate explanations, such as having a neurochemical imbalance that attracts lightning, or being cursed by a person with magical powers, or being punished by a god, or another causal explanation. However, while it is extremely unlikely to happen to any one person, it nevertheless will happen (on average) to 39 people in the absence of any other causal explanation. For those particular people, "chance alone" is an extremely difficult explanation to accept.

The basic point is that while rare events are extremely unlikely to happen to any given individual, they are almost certain to happen to some individuals within a large population – the larger the population, the more likely it will happen to someone. It can be very difficult for such a person to avoid seeking, and often believing in, some external force that has caused such a bizarre event to occur to them. Indeed, seeking an explanation is adaptive behavior, as identifying any such forces or modifiers that might exist has the potential to give a person an advantage in facilitating remarkably good (or avoiding remarkably bad) outcomes. Moreover, in some cases extremely rare events do indeed happen for a reason, and investigations of rare events can teach us a great deal. However, much damage can be done in seeking and assigning reasons to explain an occurrence that has no explanation other than chance.

This brings us back to the important issue of how to choose a financial advisor. Of all the different investment strategies that one can employ, how should one choose? As mentioned before, some wealth managers consistently beat the market, getting it right year after year. However, it should now be clear that any evidence that someone is able to predict correctly depends upon the total number of wealth managers out there making predictions (the base rate). In his book, *The Drunkard's Walk: How Randomness Rules Our Lives*, Leonard Mlodinow uses exactly this example to illustrate base rate problems. Given how many different wealth managers are out there, Mlodinow predicts that over the last 45 years there has been a 75% chance that at least one wealth manager will predict correctly 15 years in a row. Thus, even if not a single person had the ability to "beat the market," there would still be some who would have the appearance of doing so. This does not mean that there can't be someone who can beat the markets, but it certainly makes the existing evidence that one can do so less compelling and potentially not very compelling at all. Indeed, the greatest ability to make wealth may be the selling of large numbers of books explaining a nonexistent ability to predict financial markets.

BASE RATE NEGLECT IN MEDICAL DIAGNOSIS

You are worried that you may have a horrible disease, but you don't want it on your medical record, so you have an anonymous testing lab run a blood test. In the vocabulary of medical lab testing, "sensitivity" is a function of how many true positive cases of the disease will be detected by a test, whereas "specificity" is a function of how many true negatives will be picked up. The test you requested is reported to have 100% sensitivity and 95% specificity, so it seems pretty definitive.[7]

---

[7] The actual formula for sensitivity is (true positives)/(true positives + false negatives) and the formula for specificity is (true negatives)/(true negatives + false positives).

You are emotionally distraught to receive a letter from the lab that you tested positive. What you really want to know and what you need to know, is how likely it is that you really have the disease. Given the specificity of the test (95% of those who don't have the disease will test negative and only 5% of those without the disease will test positive – in other words a 5% false positive rate), it seems obvious to most people that there is a 95% chance that you actually have the illness. However, such a determination depends not only on the specificity of the test, but also on the base rate (in other words, the prevalence of the disease).

Consider that the disease in question is present in only 1 in 100,000 humans and thus is somewhat uncommon. In this case, everyone who really has the disease will test positive (due to 100% sensitivity). However, since the specificity is 95%, then 95% of people who don't have the disease will test negative and 5% will test positive. Thus, out of 100,000 people who are randomly tested, one in 100,000 will test positive (and really have the disease), and 5% of people who do not have the disease will test positive, or 0.05 × 99,999 = 5,000 people. In other words, for these test parameters and a disease prevalence (base rate) of 1 in 100,000, a positive test still only gives a 1 in 5,000 chance of having the disease.

In contrast, if the disease is common and thus present in 1 out of 20 people, then 1 out of 20 people will test positive (and really have the disease), whereas 5% of 19 people (or one person) will test positive and not have it. In other words, with a disease of a high prevalence of 1 in 20, then a patient who tests positive has a 50/50 chance of actually having the disease. Thus, the rarer a disease is, the less meaningful a positive test is, even though the specificity of the test has not changed. This shows how the base rate can substantially change the meaning of an observation.[8]

---

[8] In these examples we are using numbers for easy illustration and are rounding up fractions of people to integers.

This example exemplifies why it is dangerous to use a screening test on a broader population when a disease is very uncommon. A doctor does not typically order a diagnostic test unless there is good reason to believe the patient may have the disease; e.g., a patient has reported symptoms consistent with the disease and/or the patient has risk factors for the disease, such as an exposure history or family history. Moreover, in many cases, a thorough physical exam shows signs of the disease, which corroborate the other risk factors. Thus, by the time the test is ordered, the pretest probability (or base rate) is already quite high, and thus a positive test will have a better predictive value. This is in stark contrast to when a test is used as a general screening test for a rare disease.

Recently, guidelines for screening for breast cancer (using mammography) and for prostate cancer (using PSA) have been modified to eliminate some testing in lower risk populations. The reason for this is exactly that stated in the previous example – the lower the frequency of the illness, the larger the number of patients who will test positive who don't actually have the disease. The problem is that a false positive on the screening test can lead to invasive procedures (involving biopsy and/or organ removal) that can have serious side effects. The procedures themselves can cause pain and scarring. Infections can result from the procedures (with a low but not negligible frequency), leading to illness and possibly death in extreme cases. The process of removing the organ can cause deficits in the patient, both functional, cosmetic, and psychological. On the flip side, lives will be saved when an illness is detected and treated before it can spread. However, on a population basis, and also in the context of making the decision that is most likely to benefit an individual, one must weigh the risk of missing a real case (if one doesn't screen) against the risk of injuring, impairing, and potentially killing patients who never had the disease to begin with, but who happened to test positive. This can be a difficult calculus to analyze, and there is no objective answer on what risk is acceptable to whom, as this is a case of preference. However, unless and until the actual risk determinations are known, one cannot make an informed decision, and in this case neglect of the base rate is a serious error that is consistently made and should be guarded against.

PURPOSEFULLY MANIPULATING PEOPLE THROUGH BASE RATE NEGLECT

A very clever con makes use of base rate neglect, or in this particular case, base rate blindness. A group of grifters send out 100,000 mailings from a fake wealth management firm. Some 50,000 people receive letters predicting the stock market will rise in the next quarter, whereas a different group of 50,000 receives letters predicting the opposite. So, for half of the recipients, the prediction will come true. In the next quarter, the group sends out a second mailing to the 50,000 people who previously received a letter making the correct prediction. This time, 25,000 of the people receive a letter predicting the market will go up again, and the other 25,000 will get the opposite. In the third quarter, this is repeated for the 25,000 people who got the correct second quarter prediction (12,500 receiving a prediction of either up or down), and so on for the fourth and fifth quarter. At this point, the con artists can legitimately state to 3,125 people that they have demonstrated their ability to predict market behavior for 5 consecutive quarters. They then offer to sell these people their prediction for the next quarter at a premium price. Many people pay that price, confident about the results they have seen. But the data are compelling precisely because these people are blind to the base rate – not because they have misperceived it but because they have been purposely blinded to it. If the letter recipients knew the whole story, they would not be impressed with the ability of the economic advisors; however, not having seen the base rate, the economic advisors look very impressive indeed.[9]

---

[9] Example derived from Paulos JA. 2001. pp. 42–4.

WHAT RANDOM REALLY LOOKS LIKE

On May 23, 2009, Patricia Demauro, a grandmother from New Jersey, decided to play craps at the Borgata Hotel in Atlantic City. In craps, the first time a shooter rolls a pair of dice it's called a "come out roll." If the come out roll is a 7 or 11, the shooter wins. If the come out roll is a 2, 3, or 12, the shooter loses. However, if any other number is rolled (4, 5, 6, 8, 9, or 10), it's labeled the "point number." Once a point number is established, the shooter wins if she rolls the point number again before rolling a 7 or 11. However, any other number rolled (other than the point number 7 or 11) pays money to people who have bet on that number and the shooter keeps rolling. Thus, a shooter can keep shooting and winning money for him- or herself or the other betters (depending on what numbers were bet on), so long as neither the point number nor 7 or 11 is rolled. In the case of Patricia Demauro, her come out roll resulted in a point number of 8. She then proceeded to keep rolling the dice, not rolling an 8, 7, or 11 for 154 rolls, spanning 4 hours and 18 minutes. Was there something special about Patricia? Was she cheating, did this happen by chance alone, or is there some other explanation? Is it even worth seeking an explanation?

For simplicity, let us use the iconic example of a coin that is heads on one side and tails on the other. In addition, we shall make the assumption that the coin is a "fair coin," which is to say there is exactly a 50% chance that the coin will land heads up and a 50% chance that the coin will land tails up on any given toss. Which of the following outcomes is most likely for a series of 10 tosses? Please give this question some serious thought before coming to an answer.

1. HTHTHTHTHT
2. HTTHTHTHHT
3. HHHHHTTTTT
4. HHHHHHHHHH

To many people, choice 3 and especially choice 4 seem highly unlikely for a fair coin. Choice 1 looks about right, although the repeated pattern from heads to tails seems more ordered than a truly random system. Indeed, choice 2 is the only choice that has no obvious pattern to it, and thus seems to be the most likely pattern when flipping a fair coin.

In actuality, all of the above examples have *exactly* the same likelihood of occurring any 10 times you flip a fair coin. The chance of flipping a head or a tail is 1 in 2. The combined chance of two events happening is derived by multiplying their individual probabilities; in other words, the chance of getting two heads is $\frac{1}{2} \times \frac{1}{2} = \frac{1}{4}$. Thus the chance of getting 10 heads in a row is $\frac{1}{2} \times \frac{1}{2} \times \frac{1}{2} \times \frac{1}{2} \times \frac{1}{2} \times \frac{1}{2} \times \frac{1}{2} \times \frac{1}{2} \times \frac{1}{2} \times \frac{1}{2} = 1/1{,}024$. So, on average, once every 1,024 times you flip a coin 10 times you will get HHHHHHHHHH. This does not mean that this is guaranteed to happen, but the odds are that this will occur. The important thing to recognize is that every flip is independent of all the other flips. What this means is that there is a 1 in 1,024 chance that any of the above patterns will occur, and each of them has an equal chance.

The reason that choice 2 looks best is that when you flip coins there are a huge number of different variants of H or T that roughly look kind of like choice 2, whereas there are only two possibilities that look like choice 3 (first 5 H, second 5 T; or first 5 T, second 5 H). Likewise, there are only two possibilities that look like number 4 (all H or all T). Thus, coin flips that look like choice 2 will occur much more commonly than coin flips that look like choices 3 or 4, and the number of combinations that add up to what looks like a random sampling far exceeds those that look like an ordered sequence; however, any particular combination will occur, on average, 1 time out of every 1,024 times the coin is flipped 10 times.

Of absolutely essential importance in our analysis is the realization that although sequences that have the appearance of nonrandomness occur less frequently, they do in fact occur. From a probabilistic sense, it isn't that they might occur but that they will occur, and inevitably so. Suppose that you are given 32 coins and asked to evaluate whether they are fair coins. In this particular example, let us assume that the coins are indeed all fair. Each coin is evaluated with five flips. Even if every coin were fair, the odds of every flip being the same is $\frac{1}{2} \times \frac{1}{2} \times \frac{1}{2} \times \frac{1}{2} \times \frac{1}{2} = 1/32$. Since there are 32 coins, that means that for one of coins tested, all five flips would result in heads, and for one of the coins tested, all five flips would result in all tails. Thus, two of the coins tested, although being normal, fair coins, would (by chance alone) have the exact same outcome as when using a profoundly unfair coin that was weighted to always land heads up or tails up.

In the previous example, a person with a reasonable understanding of probability would not suspect an unfair coin if one coin came up as all H and another as all T, given the number of coins tested and the above probability distributions. However, now consider a scenario in which 32 different people were each given a single coin to evaluate, with no knowledge of the other 31 participants in the study. By chance alone, one predicts that two of these participants who received a single coin would have it come up as five consecutive H or five consecutive T after five flips. To these participants, this would seem unlikely, and they may suspect that they have an unfair coin. If they had actually lost money on each bet and didn't have the chance to test the coin any further, they would probably feel cheated by a scam of some sort. This scenario is analogous to how most of us experience life – we see only the results of a single series of events and are unaware of how often they occur in a broader context, that is, how many other flippers there are.

The bottom line is that unlikely events will happen. The more unlikely they are, the less frequently they will occur, but they will eventually occur (albeit to a very few individuals) – not might happen, but will happen. It is very difficult to dissuade people from the notion that it happened for a reason. But a reason isn't necessarily justified. Sometimes there is a reason; however, in the absence of any reason the event would still happen eventually to someone. So what about Patricia Demauro. Was there some cause that made her so lucky, or did she somehow have "skill" shooting dice, or is it simply that if enough people play over time then this will occur by chance. We never hear reports of all the people who shoot craps and lose after a few roles.

THE CLUSTERING ILLUSION

The clustering illusion is a real and serious result of poorly understanding what a random system looks like. It is a well described process by which the human eye is drawn to the clustering of outcomes as a phenomenon that suggests an underlying mechanism or cause, when in fact there is none. This issue and its related effects are expertly described by Thomas Gilovich in his book, *How We Know What Isn't So: The Fallibility of Human Reason in Everyday Life*. In this work, Dr. Gilovich gives a synopsis of the primary research carried out by him and his colleagues on the clustering illusion. The excellent example he puts forth is having "hot hands" in professional basketball – the belief that players have streaks of making shots when they're hot and streaks of missing shots when they've gone cold. A common explanation for the hot hands effect is that making a shot results in increased confidence and relaxation, giving the next shot a higher probability of hitting its mark. Conversely, missing a shot or two can rattle a player's confidence; they get uptight and have a lower chance of making the next shot. Streak shooting, then, is the result of the shooter's condition – they can "get in the groove" or "get cold."

This brings us back to the discussion of the fair coin. Neither Dr. Gilovich nor anyone else is questioning that streak shooting occurs; the data clearly show runs of consecutive shots made and runs of consecutive shots missed. Thus, the players and the fans are correctly observing the sequence of events before them and correctly identifying a pattern. I have yet to meet a person who plays basketball who hasn't experienced exactly the same thing,[10] and this extends to other sports such as golf, tennis, etc.[11] However, one needs to be mindful of the fair coin example – even in a random world, runs of

---

[10] I may be the only human who has played basketball and who has not experienced a streak of making baskets; as per my best recollection, I have never made a shot even once. However, this may be a false memory derived from a reconstructed experience. Nevertheless, I have certainly experienced going cold for basically my entire athletic life.

[11] This also extends to the lab, to experiments "working" or being interpretable.

H or T will occur with a certain frequency, just as runs of making basketball shots and missing basketball shots will occur by chance alone. This reasoning neither argues for nor against the hot hands phenomenon; however, it does underscore the need to test the hot hands hypothesis, because streak shooting will occur regardless of whether the hot hands phenomenon exists or not.

Most people agree that the hot hands explanation of streak shooting predicts that the probability of making a shot increases if the previous shot went in. Conversely, the probability of making a shot decreases if the previous shot was a miss. This is the basis of the hot hands phenomenon, because confidence, relaxation, and getting in the groove makes it more likely you will hit a basket if you just made several, whereas getting uptight and nervous after missing a few will make you less likely to make the next one. After detailed analysis of highly accurate basketball statistics, Gilovich, Vallone, and Tversky demonstrated that there was no increased probability of making a shot if the previous shot (or shots) were made, nor was there a decreased probability of making a shot if the previous shot (or shots) were missed.[12]

These findings appear to reject the hot hands hypothesis; however, as is done in science and normal human thinking (see Chapter 3), the hypothesis was rescued by someone who evoked the auxiliary hypothesis that once a player starts making shots, they are guarded more aggressively by the other team and they also may start attempting harder shots, so the lack of increased probability was hidden by these other factors. To address this, Gilovich and colleagues carried out the same analysis for free throw shots (where the shots are the same each time and the other team is not guarding the player). The same result was observed: There was no probabilistic indication of the hot hands phenomenon.

[12] Gilovich T, Vallone R, Tversky A. 1985. "The Hot Hand in Basketball: On the Misperception of Random Sequences." *Cognitive Psychology* **17**: 295–314.

To save the hot hands hypothesis, it was modified by changing the definition of the hot hands phenomenon. Advocates of the hypothesis stated that streak shooting did not exceed the frequency that is predicted by chance alone, but ballplayers had the ability to predict when they would make or miss a shot that was better than random guessing, based on their feelings of when they go hot or cold. However, formally testing this with a group of college basketball players showed no ability to predict the next shot better than chance alone. Thus, in aggregate, these data make a strong case for rejecting the hypothesis that the hot hands phenomenon exists; rather, it represents a misperceived association.

When confronted with these data, athletes are very hesitant to reject the existence of something they have clearly perceived. Students in my critical thinking class will often acknowledge and accept all of the data presented by Gilovich and colleagues, but will then reject the conclusions to which the data point, without presenting any flaw in the arguments, auxiliary hypotheses, or alternate explanations. It just doesn't sit well with them because they themselves have experienced the very illusion that is being analyzed, and "seeing is believing." Or at least "perceiving is believing," even if the perception is incorrect. It isn't that the hot hands explanation doesn't make sense. It is a perfectly reasonable prediction that, all things being equal, more confident players make more shots; however, this just doesn't appear to be the case. Please keep in mind, no one is saying that better players don't make a greater percentage of their shots than worse players do. So, better players will certainly have more streaks of shots made than will worse players. Likewise, no one is saying that how a player performs cannot not be affected by emotions or psychology with regards to the overall percentage of shots they make. However, what is being said is that for a given player that makes a percentage of shots, streak shooting within these percentages is a random clustering illusion and not that the player is actually going hot or cold. The bottom line is that we are quite capable of being susceptible to the illusion of association based on our systems of perception.

REGRESSION TO THE MEAN

In 1885, a paper was published by Sir Francis Gallon entitled "Regression towards Mediocrity in Hereditary Stature." In this work, Gallon commented that when two humans of extreme stature reproduced (e.g., both parents were very tall), the offspring were, on average, shorter than either parent. In the opening of his discussion, Gallon referred to observations he made in 1877 of the same phenomenon with respect to plant seeds. He had planted seeds of extreme size (either very large or small) and then measured the subsequently produced seeds generated by the plants that grew. The very large seeds tended to produce plants with seeds smaller than the parental seed, whereas very small seeds tended to product plants with seeds larger than the parental seed. This is not to say that the seeds derived from large seeds weren't larger than the seeds derived from small seeds, as seed size is hereditary. Rather, Gallon's observation was this: If one chooses the extremes of a given characteristic (i.e., large or small), the offspring will not maintain the extreme to the same extent; rather, they will "regress toward the mean" of the population.[13]

The effects of the regression to the mean on our perception of associations was elegantly demonstrated by P. E. Schaffner in 1985. Participants were in charge of a computer-simulated classroom; there weren't any real students. On any particular day, the "students" would be early or late for class, and the participants in the experiment could reward or punish them, and then reward or punish them again the next day based on being late or early. The students were preprogramed to come in late or early on any particular day, and, unknown to the participants, the reward or punishment thus had no effect whatsoever on the students' behavior.

Predictably, the participants tended to reward when the behavior moved in the desired direction (earlier) and tended to punish when

---

[13] A detailed description of regression to the mean is excellently described in Gilovich's book, *How We Know What Isn't So*, and several of the following examples presented here are likewise presented there in excellent detail.

the behavior went the opposite way (later). Approximately 70% of participants reported that punishing was more effective than rewarding. The informative point here is that the behavior the computer was reporting was preprogramed and entirely independent of the actions of the participants. Therefore, the participants perceived an advantage to punishment when there was none, as the subjects' actions were predetermined and could not be affected. Hence, a strong perception of an association was obtained (that punishment improved the behavior of the fictional subjects) when it wasn't possible for that association to exist.

Why did the subjects in this study have the perceptions that they did? Assuming a bell-shaped curve type distribution, then on average, really late students are more likely to come in earlier the next day just by chance alone, and the more extreme their lateness on the previous day, the more likely they are to arrive earlier on the next day than their previous arrival. For example, there is the latest that one will ever arrive and the earliest one will ever arrive (the extremes). One is much more likely to arrive earlier the next day after having arrived the latest that one ever will the previous day. The closer to the extreme, the more likely the next day will be moving toward the middle, and hence the term regression to the mean.

One typically only punishes students when they are late, and one is more likely to punish them the more extreme their tardiness. Thus, punishing has the appearance of causing the students to arrive earlier on subsequent days. In contrast, one only rewards students when they are early, and on average, the earlier a student arrives, the more likely they are to arrive less early on subsequent days – the opposite trend of what is desired. Thus, even for a predetermined pattern, regression to the mean (the extremes tending to be more average going forward) gives a very clear (but false) impression of cause and effect regarding modifying any behavior.

The importance of the notion of regression to the mean, is that people who observe this regression attribute it to some underlying association that doesn't exist. It isn't that people are misperceiving the pattern in the data. On the contrary, they are observing things very accurately: An extreme is being followed by a less extreme event. The problem lies in a lack of understanding that regression toward the mean is exactly what one should expect in a completely random system in which there are no associations between intervention and outcome. This doesn't mean that an association can't occur when data regress to the mean, but it does indicate that an association need not occur to explain the regression. Indeed, it is a common perception of parents that corporal punishment results in better behaved children, when in fact the opposite is likely true. Corporal punishment actually results in maladjustment and aggression; however, spanking and other forms of corporal punishment appear to work (in part) due to a regression to the mean misperception as in the previous example. Children behave better after having been spanked, but they would have behaved better anyway even without having been spanked, as the behavior in question was an extreme that will be followed by less extreme behavior. It is important to note that parents are actually "observing" that behavior gets better after spanking, they *see* a real effect. This is one reason why parents educated to the negative effects of corporal punishment may persist in using it, it has the strong appearance of working, when it does not.[14] Many people put their personal experience over more analytic studies precisely because the outcome of the studies are at odds with what people have perceived. The notion of "seeing is believing" falls into the trap that seeing (and the interpretation that follows) not only can be flawed, but when done by humans, *is* flawed and in very predictable and consistent ways.

Regression to the mean is the explanation for the so-called *Sports Illustrated* curse – the belief that an athlete's career will be cursed if his or her picture ends up on the cover of the magazine.[15] *Sports Illustrated* only puts on its cover athletes who have had

---

[14] Durrant J, Ensom R. 2012. "Physical Punishment of Children: Lessons from 20 Years of Research." *Canadian Medical Association Journal* **184**(12): 1373–7.

[15] This example is also presented with excellent detail in Gilovich T, *How We Know What Isn't So.*

exceptional performances (e.g., recently won a number of major tours, championships, gold medals, etc.). No one has exceptional performances that exceed their average talent for long. The odds are great that after being on the cover of *Sports Illustrated* an athlete's subsequent performance will get worse, because the cover selects those who are performing exceptionally well. This is not to say that the skill of athletes featured on the cover of *Sports Illustrated* aren't better than most other athletes, but it does indicate that people are more likely to be on the cover when they perform at the height of their potential, and are thus likely (on average) to decrease in performance afterward. Conversely, imagine if there was a sports periodical called *Choke* that featured great athletes who did much worse than expected based on their typical performance. A superstition would likely arise that if you could get your picture on the cover of *Choke*, it would be really good luck – you would snap your slump and perform well again. The poor performance of an athlete would receive all the attention, and just by chance alone he or she would do better moving forward, on average. Superstitious athletes would be clamoring to get on the cover.

The regression to the mean fallacy can rear its head in all manner of personal experience and scientific research. For example, patients afflicted with illness tend to seek medical treatment when their symptoms are most extreme. Thus, regression to the mean can make many therapies seem efficacious, when in fact the symptoms would have gotten better on their own. However, the therapy appears to work precisely because the patient receives treatment (by choice) only when the symptoms are their worst. This example has shown up even in controlled experimental trials; patients addicted to alcohol show improvement in the control arm of clinical trials, simply for having enrolled in the study. It is for this reason that private education companies who sell classes to help performance on standardized tests can be very confident in their product. Students who do very well on SAT tests don't take the test again. In contrast, students who do poorly often seek help from "experts" who can teach them how to take the test better. Arguably, the worse a student does, the more likely that student is to seek help. However, the worse a student does, the more likely that student will do better the next time simply by regression to the mean. This is not meant to imply that SAT preparation classes don't lead to improved scores, but the consumers of such a product would (on average) have better scores on retaking the test whether the classes worked or not.[16]

REAL WORLD DANGERS OF FALSE PATTERNS COMBINED WITH REGRESSION TO THE MEAN

The *Sports Illustrated* curse is relatively harmless; there are more serious real-world consequences in misunderstanding regression to the mean. Base rate neglect, the lottery fallacy, failure to perceive what random should look like, and the clustering illusion can all lead people to notice "trends" that don't reflect any real underlying association. When people then attempt to take action to change the situation, regression to the mean phenomenon can make it look like the actions remedied the nonexistent effect, reinforcing the incorrect perception of both cause and effect. For an example of this, we again turn to the excellent work of Thomas Gilovich.

While he was visiting the northern part of Israel, a flurry of deaths occurred – basically a clustering of people passing away over a short period of time. There was no indication that these deaths were not natural, but they nevertheless caught the attention of people. A group of rabbis decided that these people were dying as a result of divine punishment for the sacrilegious act of allowing women to attend funerals (forbidden under some types of traditional Jewish law). The rabbis issued a decree banning women from attending any more funerals. Sure enough, right after this decree was announced, the death rate dropped to a more normal level.

As Gilovich points out, examples like this illustrate how the misperception of random sequences and the misinterpretation of

[16] Gilovich T. 1991. *How We Know What Isn't So: The Fallibility of Human Reason in Everyday Life.* New York: The Free Press.

regression can lead to the formation of superstitious beliefs. Furthermore, these beliefs and how they are accounted for do not remain isolated convictions, but serve to bolster or create more general beliefs, in this case, about the wisdom of religious officials, the "proper" role of women in society, and even the existence of a "powerful and watchful god." It is the same thinking that has led to the torture and execution of innocent individuals based on some supernatural explanation of random unfortunate events. It is an enlightened view that will acknowledge these effects and be on the lookout for them, because commonsense thinking would have us embrace them, without question, because they are so obvious to human perceptions and therefore must be true.

8 The Association of Ideas and Causes, or How Science Figures Out What Causes What

> Humans are so intelligent that they feel impelled to invent theories to account for what happens in the world. Unfortunately, they are not quite intelligent enough, in most cases, to find correct explanations. So that when they act on their theories, they behave very often like lunatics.[1]

– Aldous Huxley

OBSERVATION OF ASSOCIATIONS AS A SOURCE OF BELIEFS

So far, we've discussed how common it is for humans to misperceive individual events or groupings of random occurrences. However, a higher level of complexity can occur when one is assessing causal associations; i.e., one thing that appears to cause another.

A famous experiment was described in 1947 by Dr. Burrhus Frederic Skinner, inventor of the operant conditioning chamber (i.e., the "Skinner Box"). Dr. Skinner placed pigeons into a controlled environment where he could provide them with an agreeable stimulus (in this case, food) and observed their behavior. It is a well-appreciated principle of conditioning that many animals can pick up on associations. That is to say, if one gives an animal a reward (usually food) each time it performs a certain action, the animal will learn the association and will repeatedly perform the task in order to obtain the reward (the basis for much of animal training, not to mention for conditioning human behavior).

In the course of his research, Dr. Skinner asked a somewhat different question: How do animals respond to their environment

---

[1] This passage was edited from the original to make the language gender neutral.

when no predictor of outcome is present; i.e., how do they behave when there is no association present? To study this, he placed pigeons in a box and gave them food at regular intervals regardless of the behavior they exhibited.[2] In other words, there was no behavior they could possibly exhibit that would increase or decrease the rate at which they got food, a situation in which there was no association to be discovered between behavior and reward. The question being asked was this: Would the pigeons find associations where there were none and act on them? The answer Skinner found was a resounding yes.

Much like many humans, pigeons are not still and sedentary creatures; at any time when they might randomly receive food, a pigeon is likely to be engaging in some kind of movement or behavior. A number of the pigeons in Dr. Skinner's experiment made an association with the behavior they just happened to be exhibiting prior to the food arriving and began repeating that particular behavior over and over again. When subsequent food arrived while they happened to still be engaged in the behavior, this reinforced the perceived association and perpetuated the behavior, even though the timing of food arrival was predetermined and unrelated to the behavior. Some of the observed behaviors included head thrusting, head lifting, and turning counterclockwise. Perhaps Skinner's most striking observation was that if food was no longer given to the pigeons, they would continue to perform the behavior for a long time before giving up on it. It is unlikely, given their small cerebral capacity, that the pigeons cognitively formulated a reasoned belief that a given behavior helped them get food; nevertheless, they acted in a manner consistent with having formed a belief, which Skinner likened to a "superstition."[3]

---

[2] Skinner BF. 1948. "'Superstition' in the Pigeon." *Journal of Experimental Psychology* **38**: 168–72. Reprinted in 1992 in *Journal of Experimental Psychology* 121(3): 273–4.

[3] Michael Shermer gives an excellent description of Skinner's work as well as Ono's work in his book, *The Believing Brain,* as well as far more detail on human psychology of association. Primary references are provided, and although the work is described in my own words and with some difference in detail here, I fully acknowledge Dr. Shermer's work and influence.

Of course, pigeons do not have our "big brains," so does this model have any relevance to human behavior? A number of different studies have been carried out to test whether humans behave in a similar fashion. One particularity informative study was reported in 1987 by Dr. Koichi Ono, who reproduced Dr. Skinner's experiment using a version of the Skinner box that could be applied to humans. People were seated comfortably before three separate levers. A reward was provided to the subjects at a variable frequency, but unrelated to any behavior (in this case, food was replaced with the reward of a flashing light and ringer linked to a point counter). The subjects were told to try and get as many points as possible, but they weren't told whether any action in particular would necessarily get them points. Many of the subjects started messing with the levers (although they had not been specifically instructed to do so). As with Dr. Skinner's pigeons, the subjects were likely to have done something with the levers prior to a point registering on the counter; also like the pigeons, the subjects started repeating whatever they happened to be doing just before the point was awarded; e.g., if they pulled the levers in a particular order or with a certain timing. Some people touched the counter, touched other objects, or jumped up and down in certain patterns. In each case, the behavior initially preceded a reward and was subsequently reinforced if the person happened to repeat it until another point was awarded. The incorrect perception was that the behavior caused the reward, when in fact the reward was a random occurrence.

The subjects in Dr. Ono's studies demonstrated that in the absence of any existing behavior that can cause a desired outcome, humans nevertheless have the tendency to seize on coincidental occurrences and to become increasingly convinced that causal associations actually exist. The more they believe this, the more they perform the behavior, and the more the belief is reinforced, if and when the result occurs again. It is argued that this is where superstitions come from. This is not an example of misperceiving an actual association as causal when it is not, but rather an example of generating a strong belief in a causal association where no association (not even a correlative one) exists.

Together, these observations place human actions and rituals believed to be beneficial in a hazy light. To be sure, humans have discovered many real and causal associations, giving us an ability to alter our environment to a degree that far exceeds any other animal on Earth; however, the undesired side effect of this open-minded search for associations is the accumulation of many false beliefs based on nonexistent associations.[4]

Superstitions are generally ubiquitous in humans and are both personal and cultural. We may obtain new superstitions and lose old ones as we progress from childhood into adulthood and into old age; however, we maintain a general set of beliefs around a set of actions (or avoidance thereof) that may lead to desirable outcomes, based on ongoing experience. This is found perhaps most famously in the sporting world, and in baseball in particular, where players have a wide variety of ritualistic "good luck" behaviors when coming up to bat. Certainly many gamblers also have good luck rituals and special "lucky" objects. Even our political leaders have strange associative beliefs.

Tony Blair always wore a certain pair of lucky shoes when addressing Parliament,[5] John McCain carried a lucky feather, penny, and rock,[6] and President Obama always played basketball on Election Day, as well as carried "a lucky American eagle pin, a small image of Madonna and child, and a tiny figure of a Hindu monkey god."[7] One

---

[4] These examples are adapted from an excellent treatment of this topic in Shermer M. 2011.*The Believing Brain from Ghosts and Gods to Politics and Conspiracies – How We Construct Beliefs and Reinforce Them as Truths*. New York: Time Books, Henry Holt and Co.

[5] June 24, 2007. "Blair wears same shoes for 10 years." *ABC News*. www.abc.net.au/news/2007-06-22/blair-wears-same-shoes-for-10-years/77290

[6] Milbank D. Feb. 19, 2000. "A candidate's lucky charms." *Washington Post*. www.washingtonpost.com/wp-srv/WPcap/2000-02/19/067r-021900-idx.html

[7] Katz C. Nov. 4, 2008. "Superstition rules the day; both Obama and McCain count on good-luck rituals." *Daily News*. www.nydailynews.com/news/politics/superstition-rules-day-obama-mccain-count-good-luck-rituals-article-1.333556

might argue that such behaviors are playful actions by educated individuals who don't really believe in any connection between what they are doing and the outcome. However, it is also a fair speculation that at some level even bizarre superstitions really are believed and felt to be real. I can promise you that many laboratory workers (myself included) have a similar "favorite" item, such as a pipette, or instrument, or bench space, that they believe will make it more likely for them to correctly perform a certain technique or experiment. We could take the time to test the association, as good scientists do, but one seldom wants to spend the time and energy to test something like this, when just using the preferred item is much easier. In all likelihood, we are just pigeons thrusting our heads and rotating counterclockwise, hoping it will help the experiment work better.

MISTAKING TRUE CORRELATIONS FOR CAUSAL RELATIONSHIPS

Autism is a disorder in which individuals have diminished abilities to communicate, verbally and nonverbally, due to problems with brain development during early childhood. Signs of autism typically become apparent prior to 3 years of age. As brain cognition develops in children, children who fail to achieve certain milestones are identified. In some cases of autism, however, milestones are initially achieved and then individuals subsequently regress. The manifestations of autism range from mild to severe, and they do not necessarily indicate limitations; indeed, many people on the autism spectrum can be extremely intelligent (or even brilliant) and highly functional. In contrast, severe cases may be seriously challenged and unable to function independently, requiring significant lifelong support.

The cause or causes of autism remain unclear; however, the frequency of autism diagnoses has been increasing in recent decades, at a rate that has caught the attention of epidemiologists and public health officials. The urgency in identifying the causes of autism is profound; if we don't know the causes, we can't develop a rational strategy for either prevention or developing advances in treatment. In all likelihood, multiple factors lead to autism. Moreover, what we call "autism" may actually be a number of different processes with a common endpoint, or at least an outcome that appears similar across individual cases. Nevertheless, identifying causes would be an important step in addressing the problem.

Autism may be the result of genetic, environmental, and/or infectious factors – or a combination of all three, as well as others. However, a large number of people are convinced that childhood vaccinations are one such factor. The measles vaccine has come under particular scrutiny, with a focus on a mercury-based chemical (thimerosal) that was added to the vaccine as a preservative. Many parents have observed an association between their child getting vaccinated for measles and the subsequent development of autism. In most, if not all of these cases, the child was healthy and developing normally, received a vaccine, and soon thereafter (sometimes within days) exhibited the first signs of autism – a suspicious and compelling story indeed. Given the number of parents who have reported that their children developed autism after receiving a vaccine, this association must be taken very seriously.

The stakes in this situation couldn't be higher. If vaccination truly causes autism, then a huge number of people are being injured by vaccination practices. However, decreasing vaccination will lead to small outbreaks of measles, occurring essentially every year, with occasional large outbreaks (for example, the outbreak of 667 cases of measles in 27 states in 2014 and at least 349 cases in 26 states in 2018).[8] Measles is by no means a benign disease. Although most children with measles suffer transient illness with fever, rash, and discomfort, 28% of children younger than 5 years of age need to be hospitalized, and in extreme cases some patients die. Moreover, patients who survive serious cases can wind up with deafness and

[8] Centers for Disease Control and Prevention. Jan. 25, 2019. "Measles cases and outbreaks." www.cdc.gov/measles/cases-outbreaks.html. As of this writing, the 2018 outbreak is still underway with final numbers not yet determined.

lifelong brain damage. Importantly, young infants cannot be vaccinated and are thus highly susceptible to measles infection. In addition, people with compromised immune systems, some severely compromised, for whom infections (such as measles) are horrific and often lethal, are put at risk when other people are not vaccinated and outbreaks occur. Finally, in epidemic areas, measles remains a plague on humanity. Prior to a widespread available vaccine in 1980, it is estimated that 2.6 million humans died from measles each year. Today, despite an available vaccine, lack of worldwide distribution is such that approximately 100,000 people still die each year from measles. Thus, ceasing vaccination or limiting its distribution is not a benign maneuver and should not be taken lightly.

It is for these very serious reasons that this issue is so essential to evaluate correctly. However, how does one make such a determination in a way that is best adjusted to risk and outcome? Essential to this evaluation is the determination of whether or not measles vaccination actually results in higher rates of autism.

ASSOCIATIONS BETWEEN THINGS: CORRELATION VS. CAUSALITY

Humans are extremely talented in detecting associations between things and this has given us tremendous advantages. Noticing the association between planting seeds and the subsequent growth of crops has allowed us to develop controlled agriculture. Noticing the association between certain germs and disease has allowed us to decrease (and in some cases eliminate) disease through public hygiene, the advent of antibiotics, and the generation of vaccines. However, a particular price we pay for our excellent abilities to identify associations is the propensity to notice associations even when they are not there. This propensity is not a benign thing and can cost us a great deal.

Things may be associated with each other (and therefore correlated) for a number of reasons, one of which is one thing actually causing another. The correct identification of causal events is incredibly powerful and gives the holders of such knowledge considerable ability to both predict the unobserved and also directly affect their environments. Thus, it is of little surprise that a great deal of time, energy, and human thought is focused on attempting to identify the causes of things. Indeed, a brief survey of the newspaper will find endless speculation on the causes of crime rates dropping or rising, demographics shifting, the climate changing, the economy improving or declining, and so on. To know a cause is to understand how to predict and to change an outcome.

Importantly, there are a great number of "true correlations" that have no causal relationship whatsoever. For example, lightning is almost always seen before thunder is heard. Under careful examination, this is a very strong correlation – and quite real, not just a mistaken association. In this context, it would be reasonable to posit that lightning actually causes the thunder, although such is not the case. Rather, lightning and thunder share a common cause by the same occurrence (an electrical discharge in the atmosphere). The misperception that lightning precedes the thunder is a result of light reaching our senses faster than sound. However, even if lightning actually did precede thunder in real time, but both were the result of electrical discharge, then lightning and thunder would still be correlative but not causally related. The coincidence of two variables may reach 100%, or, in other words, a perfect correlation (i.e., they may always occur together or one may always precede the other); however, as in this example with lightning and thunder, even this perfect correlation does not necessitate causality.

Another example is found in the instance where a hunter shoots a quail with a gun. To a third party who is watching, when the hunter pulls the trigger the following events occur: a loud sound is heard and a quail falls out of the sky. Now, to most people who grow up in a society in which guns are understood, it would be obvious that a bullet hit the bird. However, an alternate interpretation is that the loud sound is what caused the bird to die, perhaps by frightening it. This would be based on a 100% correlation, as every time a bird is shot out of the sky, a loud sound is heard. Moreover, the bullet itself is

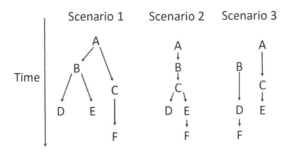

FIGURE 8.1 Correlation vs. causation

not observed, as it is too small and fast to be seen by the human eye; thus, in some ways, the loud sound theory carries much greater observational evidence. Even if a bullet is stipulated, it is certainly reasonable to assume that the loud sound caused the bullet to leave the gun; after all, the loud sound always precedes the bullet being shot. Again, to an educated person familiar with the working of guns, this seems a bit absurd; however, it is also very logical.

Consider scenario 1 in Figure 8.1. If A causes C, then A will always occur before C. However, if A also causes B (and B does or is perceived to precede C), then B will also always occur before C. If the observer can't perceive A (as someone watching a gun being fired can't see the trigger being pulled or the gunpowder in a bullet explode), then it is easy to mistake a correlation (in this case B) for the cause of C, instead of the true cause (A). B and C still have a 100% correlation, but have no causal relationship to each other.

The potential for confusion becomes even greater as one considers the further unfolding of events in time. Let's continue down Figure 8.1 and see that B causes D and E, while C causes F. Since C precedes both D and E, and C always occurs before D and E, then it is reasonable to conclude that B causes C and then C causes D and E. In other words, a passive observer cannot distinguish between scenario 1 and scenario 2, despite the causal relationships being quite different. These two very different correlative/causal scenarios are presented in Figure 8.1. Thus, correlation can be used to conclude that two events are related in some way, but causality cannot be determined merely by observing correlation.

This brings us back to the relationship between vaccination and autism. Does vaccination cause autism? There certainly is an extremely strong correlation. However, for the reasons demonstrated here, that very strong correlation cannot, on its own, be used to conclude causality. Let's compare scenarios 2 and 3 in Figure 8.1. In both scenarios, let B represent vaccination and C represent the onset of autism. In scenario 2, vaccination (B) causes autism (C), which is why autism occurs after vaccination. In scenario 3, vaccination (B) and autism (C) are entirely unrelated, but both are correlated with time, such that vaccination will precede autism. Since the vast majority of children in the developed world receive vaccines, then essentially all children who develop autism will have been vaccinated prior to the first signs of their autism. Accordingly, the correlation will be very good and have the appearance of causality, regardless of whether vaccination causes autism or not.

CONFOUNDING WITH TIME AND SEQUENCE

The natural history of autism in humans is that the early signs are first noticeable at around 18 months of age. Humans who develop autism may not only fail to progress normally, but may show certain regressions (loss of previously accomplished milestones) when the disease begins to manifest itself. The current recommendation for administering the measles vaccine is that the first dose be given between 12 and 15 months of age. The vast majority of children in the United States are vaccinated for measles. Thus, it is an absolute and temporally dictated certainty that most children who develop autism were healthy normal babies. Then, approximately 3–6 months after they got their measles shot, they began to show signs of autism.

To be clear, this correlation between vaccination and autism does indeed exist, and it is a very strong correlation. Moreover, this observation is certainly consistent with the measles vaccine causing autism. Nevertheless, the correlation in no way demonstrates that autism is caused by measles vaccinations, as opposed to just being an association. An important consideration is to acknowledge that because of the time windows of the two events (vaccination and autism), an association (or correlation) is guaranteed to occur whether or not there is any causality at all. The point here is that assignment of causality based on correlation is equally consistent with two variables being associated (with no causal relationship) and with actual causality. However, suspecting, or even concluding causality in such situations is the normal way that humans assess the world.

There is a very strong correlation between high blood pressure (hypertension) and having a heart attack or stroke; indeed, current medical wisdom considers hypertension to be a causal risk factor in heart attack and stroke. However, there is also a very high correlation between one's hair going gray and having a heart attack or stroke. Indeed, on average, the more gray hair one has, the more likely the person is to also have a chronic illness. It would thus be a very reasonable supposition that gray hair causes many health problems.[9] While most people would likely acknowledge that the association of graying hair with health problems is that they are both caused by a common process (getting older), its appearance is no different than that of high blood pressure, which is also often a result of aging and is considered a health problem. Because aging is associated with poorer health, then anything else that is also associated with aging will be associated with poorer health. This is called a "confounder," and it has reared its ugly head many times, fooling all manner of people, including very good scientists.

---

[9] It should be noted that much additional evidence exists to implicate hypertension in causing heart attack and stroke, and that lowering blood pressure to normal levels greatly decreases heart attacks and strokes; to the best of the author's knowledge, no one has ever formally tested the theory that providing hair dye to everyone with graying hair would save lives through the prevention of such diseases.

THE APPEARANCE OF CAUSALITY

The very nature of causality and the consideration of whether causality actually exists at all, has a substantial history of thought in the philosophical arena. To the nonphilosopher, the deepest philosophical conundrums may at times seem like analytic silliness carried to an extreme. At first glance, one could easily reach such a conclusion when evaluating the volumes of philosophy and the lifetimes of thought addressing the issue of causality. In general speech and perspective, this is simple and noncontroversial. For example, few people would dispute that when I turn on the light switch I caused the light to come on. However, the issue of causal relationships, especially in the debate about how deterministic the world really is, is a complicated and fascinating quagmire of which the reader should be aware (although we will not walk into that quagmire in this work). What is highly relevant to this work is the notion, raised by the regrettably correct David Hume and many great scholars who followed, that one cannot observe causality. That no human has ever observed causality is a notion of fundamental importance.

To use one of David Hume's examples, when one billiard ball hits a second ball, the second ball moves. It appears as though we have observed causality. However, careful analysis reveals that in fact we have not observed causality at all. Based on our experience of the world it seems causal, but the fact of the matter is that we have only observed one thing occurring before another. There is no special property of causality that we can observe; if causality is observable, our senses can't pick it up. The best we can accomplish is to observe temporal relationships; in other words, one thing occurring before or after another thing. If I shoot a gun at someone who then dies, an observer wouldn't see the gunshot causing his death. That person would observe two temporal incidents: I shot a gun, a person then died.

The British philosopher John Stuart Mill gave special attention to this issue in his seminal work, *A System of Logic, Ratiocinative, and Inductive*, first published in 1843. He stated the problem of observing causation as follows: "Observation, in short, without experiment (supposing no aid from deduction) can ascertain sequences and co-existences, but cannot prove causation." An example provided by Mill that illustrates the point quite nicely is that of day and night. In human experience thus far, no day has ever been observed that wasn't preceded by night. One might call this a tautological definition, as day and night define each other. Nevertheless, one could conclude that night always precedes the day; therefore, night must cause the day. Likewise, the day must cause the night. Hence, one could think of day and night as causally related, where each one is the cause of the other. However, our current understanding is that both are caused by the rotation of Earth with regards to a relatively stationary Sun; thus, their perfect association is the result of each being caused by the same thing, not because one causes the other. Nevertheless, knowing nothing about Earth's rotation, the observation is equally consistent with day causing night and vice versa, and there is no reason that such couldn't have been the case. Thus, as discussed earlier, even a 100% association of one thing always preceding another is not evidence of cause, just association. Hume's point is a profound one; we assign causality, but we have no direct evidence of causality, as we do not observe causes, only associations.

ASSESSMENT OF CAUSALITY: THE ISOLATION OF VARIABLES

So, if one cannot observe causality, how can one ever assess causality? It is from this central question that much of the theory and practice of scientific method is derived. If scientists retroduce hypotheses about the causes of observed effects and we wish to test such hypotheses, then in the absence of some ability to observe causation we must find a way to test causation. If one cannot observe causality as a property, then how can we ever assess whether one thing actually causes another?

How can we test if HIV actually causes AIDS, if smoking really causes lung cancer, or if vaccinations really cause autism? Again we can turn to John Stuart Mill, who has addressed this by defining a number of rules about making determinations in this context. At the end of the day, Mill and others have basically come to the following point of view. One cannot test causality from a single occurrence; rather, one must compare at least two separate occurrences that differ by one and only one thing, and see whether it changes an effect.

For example, consider two situations in which everything is identical, every possible variable and factor is identical, but with one and only one difference. Let's say you have two houses that are identical in all possible ways. In both houses, the light switch in the bedroom is off and the lightbulb is not emitting light. You then turn on the bedroom light switch in one house (but make no other changes in the house whatsoever) and the lightbulb emits light. In this case, one could conclude that turning the light switch to the "on" position was necessary for the light to come on, and that turning the light switch on "caused" the light to come on. One still hasn't observed causality, but one has *reasoned* causality based on the comparison of the two situations; if indeed the position of the switch was the one and only difference, this reasoning is valid. It is precisely the same reasoning employed in the early example of changing the battery of your car when it wouldn't start. If the battery is the only thing you changed and the car started, then one concludes that the dead battery was the "cause" of the car not starting.

Mill defined a number of other logical scenarios that one could use to isolate cause, but they all depend upon relatively simple systems with the ability to obtain scenarios such as the two houses example (e.g., all things are the same or different, or a combination of the two, except for a single variable).

ASSESSMENT OF CAUSALITY AS A DEMARCATION OF SCIENTIFIC PRACTICE

One can never guarantee a scenario in which two things are identical but for a single difference, and so one can never truly achieve Mill's ideal for assessing causality; however, it is an attempt at the ideal that is the logical underpinning of much scientific practice. In many cases, the purpose of a scientific laboratory is an attempt to control all the variables between two situations, and to alter only one variable. This can be achieved in a number of ways, but, most simply, one can do either of two things: (1) take two identical situations and add a new thing to one but not the other or (2) take two identical situations and remove one thing from one but not the other. If one wants to know if a drug has an effect, take two identical groups and give the drug to only one group. If one wants to know if a gene causes a disease, take two identical groups and remove that gene from only one group. In both cases, one must keep all other things equal.

Why do all other things have to be held equal? Because if more than one thing changes at a time, then one cannot tell if the effect was a result of the drug you added or a different thing. Again, this problem stems from the fact that we cannot observe a drug causing any effect; we can only observe the effect and conclude that it was due to adding the drug. If another thing changed in addition to adding the drug, which was not itself caused by the drug, we would have no way of knowing if the effect occurred from the drug or that other thing.

It is precisely this issue that makes the notion of a holistic scientific construct, which we discussed extensively in earlier chapters, such an essential consideration. Saying that only one thing has changed (and thus nothing else has changed) is basically forbidding any maneuver to maintain hypothetico-deductive (HD) coherence through modifying auxiliary hypotheses. To the credit of Quine (and others) we can never know for certain that nothing else has changed, that no change in auxiliary hypotheses is possible. However, it should now be clear why getting as close as possible to holding all auxiliary hypotheses constant is one goal of scientific experimentation – it is the only way to approach isolating a hypothesis to test or assessing causality of something. While holding all auxiliary hypotheses constant is not achievable, it is necessary to always attempt that goal and get as close as is feasible.

In practice, there are a number of ways to attempt to achieve such experimental control. In the lab, one can set up controlled, albeit contrived, scenarios to achieve this end. But while one achieves control in the lab, one also risks that the contrived situations may no longer reflect the "real world." In contrast, one can perform experiments outside of the lab in the real world and still attempt to achieve two scenarios that differ in only a single variable. An example might be a clinical trial where 1,000 patients are enrolled in a study and half of them get a new drug being tested, whereas the other half gets a placebo. In these cases, great efforts are undertaken to "randomize" the patients into either one group or the other. The word "randomize" is basically an attempt to distribute all other variables in the enrolled patients equally among the two groups, so that getting the drug or not getting the drug is the only difference between the two groups. If the group getting the drug does better than the group not getting the drug, and if indeed all other things are equal, then one can conclude that this drug has efficacy to treat the disease (or, it "causes" an improvement in the disease).

REDUNDANCY IN EXPERIMENTAL SYSTEMS

The previous section demonstrates how one can "prove"[10] a variable causes an effect, which means in some senses that it is necessary. In other words, the light switch being on is necessary for the light to be on, and turning off the switch turns off the light. However, what about the inverse proposition – can one "prove" that a variable is not

---

[10] Because of real world holism and an infinite number of auxiliary hypotheses, proof to a logical certainty is never obtainable. However, the word "prove" is used in the sciences in looser way.

necessary? If the light remains on regardless of whether the switch is on or off, then one can conclude that the switch is not necessary. However, one cannot conclude that the switch is not causally involved in the light going on, even though the position of the switch does not alter the light. The reason for this is that there may be redundancy in the job performed by the switch.

In other words, let's assume that there are two separate switches, either of which can complete the circuit to turn the light on. The other switch may be in a separate room or simply in a place we cannot access. If such is the case, then the experimental result of the light remaining on, regardless of the position of the switch we are studying, takes on a different meaning. Should one discover the other switch and turn it off, then the position of the first switch would indeed convert to a "necessary" status, as its position would then control the light. Thus, the observation that removal of variable A has no observable effect on outcome B is equally consistent with A having no role in causing B, and with A causing B but being redundant with a separate variable C that also causes B. In the case of two redundant causes, neither causal variable is technically "necessary" when the redundant variable is present, but they still contribute and may become necessary should the redundant variable fail. Regrettably, redundancy is not limited and there can be multiple redundant causes for any outcome.

Imagine the situation where two duck hunters simultaneously shoot the same duck. They shoot at the identical moment and each bullet hits the duck at the same time, delivering a lethal injury. If you remove either hunter from the scenario, the duck still dies. Often, causation is defined as the effect would not have occurred but for the cause (i.e., removing the cause removes the effect.) However, this simplistic view does not take redundancy into account. If we were to not allow redundancy to enter our thinking, then we would reach the conclusion that neither hunter caused the death of the duck, since the duck would have died even if either (but not both) of the hunters were removed. Although neither hunter is necessary, either is sufficient, and to say that neither had a causal role seems an absurd conclusion. This is why redundancy is so important. Since much of the world is invisible to us, experimental scientists face a situation where there may be endless additional hunters whom we cannot observe and of whom we are thus unaware. Again, the assumption that there are no invisible hunters is a form of auxiliary hypothesis.

SCIENTIFIC APPROACHES TO ASSESSING CAUSALITY

Let us consider the real-world issue of hormone replacement therapy for postmenopausal women. In their 50s, women typically experience menopause, which is the result of a decline in their ovarian function as they move beyond standard reproductive age. As women age, they also have an increase in osteoporosis and heart disease. It has been hypothesized that the increase in these ailments is caused by a change in hormonal balance occurring during the menopause process. If correct, then giving women back the hormones that they have lost (i.e., hormone replacement therapy, or HRT) would be a rational intervention to prevent (or at least decrease) the frequency of these diseases. After HRT became available in the 1940s, some women were given it by their physicians, whereas other women didn't receive it because their physicians didn't think it was a good therapy, they couldn't afford the therapy, or because they were not under a physician's care. In the light of this landscape, researchers began to ask the question: Does HRT actually cause a decrease in the rate of diseases that normally increase in frequency after menopause? This question, which is simple in its concept, was difficult to answer.

To study the question, the rates of osteoporosis and heart disease were compared between woman who were taking HRT and those who were not. This was a retrospective analysis by design, which means the researchers compared women who had already taken HRT vs. those who had not. Attempts were nevertheless made to control for other variables (i.e., confounders), such as the age of the woman being analyzed. As predicted from the basic rationale behind

HRT therapy, women who were taking HRT had lower levels of both osteoporosis and heart disease.[11] These data seemed compelling, supporting the HRT hypothesis, and resulted in stronger recommendations by professional medical associations that all postmenopausal women take HRT.

So, what is the problem with this retrospective study on HRT? Because it is a retrospective study we know it is more susceptible to confounders. The term *confounder*, commonly used in scientific literature, refers to the failure of the attempt to change one thing, *and only one thing*, in the system; rather, there is some other difference that may "confound" the effort to assess causality through isolation of one variable.

So, how might a confounder have worked its way into the HRT case? In order for a confounder to occur, there must be a separate variable that is correlated to the outcome being studied. Since many variables were controlled for (age of the participants, etc.), then if a confounder is present it's presumably a different variable from the ones already considered, or the variables considered thus far were not really controlled for properly. However, the results of this study were so compelling and also consistent with existing medical theory (i.e., it did not require any modification of the web of belief) that many felt it definitive and no further research was needed. However, out of this concern for hidden confounders, others proposed to run a follow-up prospective study. Prospective studies, especially those called randomized controlled trials (RCTs), are a way of ferreting out hidden confounders and adjusting conclusions accordingly. Ultimately several RCTs were run, and a hidden confounder of great importance was detected that altered the outcome.

Because HRT was generally recommended by physicians, those woman who sought out (and had access to) good healthcare were much more likely to receive HRT than those women who did not.

---

[11] Lobo RA. 2017. "Hormone-Replacement Therapy: Current Thinking. *Nature Reviews Endocrinology* 13(4): 220–31. doi:10/1038/nrendo.2016.164

This means that, as a general group, the women receiving HRT were more likely to be paying attention to other issues of healthy living (getting regular medical exams, exercising frequently, following a proper diet, testing for and treating high cholesterol, not smoking, etc.). The RCTs randomized women regardless of their HRT-seeking behaviors, thus presumably controlling for these variables. All of the women who were seeking HRT were split into two groups in a fashion that was meant to randomize the groups, so as to make all things the same between the groups other than the variable being tested (in this case, HRT). The first group received HRT, whereas the second group received a mock treatment (a placebo). In many cases, such trials took the additional caution that neither the patients nor the physicians knew what group a given patient was in. This is called a double-blinded study (because both patient and physician are "blinded" to what group a given patient is in). Double blinding in this way prevents more insidious biases, such as a physician inadvertently treating those on HRT differently than those getting a placebo, with respect to other factors (frequency of lab testing for cholesterol, etc.).

At the end of the studies, the results were striking and some-what shocking. Not only did HRT therapy not result in a decrease in the rate of heart attacks, but in one study it had the very opposite result – HRT therapy *increased* the frequency of heart attacks.[12] How was this possible? In this case, the decrease in heart attacks from all the behaviors associated with seeking HRT (better healthcare, pursuing a healthy lifestyle, etc.) was strong enough to overcome the increased rate of heart attacks from HRT, such that retrospective studies of HRT use showed the opposite result of the actual effects uncovered in the prospective study. Thus, the confounders provided not only an erroneous association, but one so strong as to overcome a real underlying association and provide the opposite result.

[12] Hulley S, Grady D, Bush T, et al. 1998. "Randomized Trial of Estrogen Plus Progestin for Secondary Prevention of Coronary Heart Disease in Postmenopausal Women. Heart and Estrogen/Progestin Replacement Study (HERS) Research Group. *Journal of the American Medical Association* **280**(7): 605–13.

Accordingly, and in light of the new data, professional medical associations reversed their previous policy and advised against HRT for most postmenopausal women.[13]

From the lay public's point of view, the entire story regarding HRT was very disconcerting. It shows the inconsistency and "flip-flopping" of scientific understanding as it evolves, and it gives a strong impression that the scientific community doesn't know what it's doing. HRT was strongly recommended as having a great known benefit, only for patients to be told a decade later that the previous recommendation not only lacked efficacy but was harmful to their health. How could such folly be the stuff of science, and how do we know that the new recommendation won't be reversed again 10 years in the future?

Well, in actuality, this example demonstrates that science is working exactly as it is supposed to work, subjecting hypotheses (even those that are broadly accepted) to increased and ongoing rigor, and reevaluating knowledge claims as more and more data become available. Although authority figures and professional societies banged the drum of HRT efficacy (admittedly based on the best data available), skeptical scientists continued to study the question, focusing on decreasing known sources of error (e.g., confounders), and generating additional data. At the end of the day the natural phenomena dictated the conclusions and not the notion of personal or group authority. Nevertheless, science is always a work in progress, "truths" are always tenuous, and no knowledge claim is immune from future revision.

The uncertainty of the scientific process is the price we pay for a self-correcting process based on increasing bodies of evidence, specifically designed to challenge previous conclusions. It is exactly this uncertainty that is the ironic virtue and strength of science. Systems

---

[13] Of note, unlike heart attacks, HRT decreased the number of fractures as a result of osteoporosis and is thus still used in some women who are susceptible to this disease. Too often Occam's Razor seems not to cut very well.

of belief that don't scrutinize their own ideas with the goal of rejecting them will never change their view, and as such have the apparent virtue of being unwavering; however, they are also incapable of correcting any errors they may have made. So, unless they are entirely perfect in all regards, there is no remedy to it – indeed, they may never even discover that they are imperfect. Being unwavering may be an essential comfort for some, who would rather a constancy than the greatest correctness over time; for others, a journey towards ever greater correctness is well worth the lack of any answers that can be held as absolute truths.

DISTINGUISHING SCIENCE BY ITS ATTEMPT TO CONTROL FOR BIAS AND CONFOUNDERS

The design of randomized double-blinded trials is essentially an attempt to compensate for confounders, both of circumstance and of human bias. Selecting a single population of people and randomly assigning them to one of two groups is an attempt to create a situation in which there is one (and only one) difference between the groups (e.g., getting HRT or not). Blinding the patients and the healthcare providers to which group was actually getting HRT is an attempt to compensate for inadvertent human biases and errors, examples of which have been discussed in previous chapters. Scientists have learned over time that such approaches to compensate for normal biases are essential to limiting erroneous conclusions. This approach is not perfect, but it is superior to situations that are less well controlled for confounders, which is precisely why the early studies on HRT produced the wrong answer.

As a general principle, neither individuals nor groups that are nonscientific carry out controlled trials to randomize groups and isolate a variable for study. There are all manner of claims made about the natural world, from the effects of herbal remedies, to the efficacy of spiritual healing, to the ability of astrology to predict the future for a given person. Many of these claims could be rigorously assessed by the same type of trials described earlier, but the makers of the claims typically do not (and often will not) do so. This is a serious demarcation between modern science and nonscience, or pseudoscience.

So, how does a scientific approach to causality help us in our determination of whether exposure to vaccines increases the chance of a child developing autism? Moreover, how does this activity fit into the processing of HD coherence and logic that we explored in the first section of this book?

First and foremost, we need to acknowledge that noticing the association between vaccination and the development of autism is a highly valid observation, and it is only appropriate that astute parents and providers should always be vigilant in their capacity to notice such trends. They are the vanguards of observing the real world and detecting new associations, and they are to be lauded in this regard. Second, it is likewise appropriate to be mindful of the fact that, as described previously, the association between vaccination and autism would occur regardless of whether a causal link existed. That is to say that in both the scenarios where vaccination did play a causal role and in scenarios where it didn't, vaccination would nevertheless appear to cause autism due to the timing of vaccination schedules for children and the natural timing of autism onset. In an HD framework, the retroduction of the hypothesis that vaccination causes an increased risk of autism is a good retroduction, as the posited notion would predict an observed effect and also lead to other testable predictions. However, this is where most normal human thinking would stop, which is to say, using particular approaches to rigorously test a notion outside the context of normal human experience is not typical human behavior; rather, this is an activity more commonly associated with scientific practice.

What are some testable predictions of the hypothesis that vaccines cause an increased risk of autism? If the measles vaccine causes autism, there should be an association (timewise) between vaccination and autism. A study published in 1998 by Dr. Andrew Wakefield and colleagues studied 12 children with autism.[14] In eight of the children, the signs of autism were noted to have started after a measles vaccination; however, this is just codifying what is, as explained previously, a real and inevitable temporal association. In addition, it was claimed that the measles virus was recovered from the intestines of the afflicted children, suggesting a mechanism for the vaccine effects. This study caused a near panic, resulting in substantial declines in measles vaccinations and fueling fears that vaccinations in general increased autism (through increased activation of the immune system), that measles vaccines in particular caused autism, and that thimerosal (included as a preservative) had a major role in causing autism. This led to a tremendous amount of research on the issue.

If the idea is that vaccines increase the risk of autism by exposing the immune systems of children to more and more foreign things, then the more vaccines to which a child is exposed, the higher the rates of autism should be. This prediction follows from the general hypothesis that exposure to more foreign things increases autism. To test this idea, a study was performed. The exposure to vaccines was compared in 256 children with autism and 752 children who did not have autism, trying to keep all other things equal between the groups (age, male vs. female, managed care organization, etc.). No difference in the extent of vaccination was observed between the groups.[15] A similar study was carried out on the preservative thimerosal. If exposure to thimerosal increases rates of autism, then there should be a greater rate of autism in children who have greater exposure to thimerosal (all other things being equal); however, this was not

[14] Wakefield AJ, Murch SH, Anthony A, et al. 1998. "Illeal-Lymphoid-Nodular Hyperplasia, Non-Specific Colitis, and Pervasive Developmental Disorder in Children. *Lancet* **351**(9103): 637–41.

[15] DeStefano F, Price CS, Weintraub ES. 2013. "Increasing Exposure to Antibody-Stimulating Proteins and Polysaccharides in Vaccines Is Not Associated with Risk of Autism. *Journal of Pediatrics* **163**(2): 561–7.

observed.[16] In observational studies, autism increased during the 1980s and 1990s both in countries that used thimerosal-containing vaccines and those that discontinued its use due to concerns surrounding an autism link; autism rates continued to rise in both populations.[17] Thus, many outcomes that could be predicted from the thimerosal hypothesis were not observed. In all fairness, it's not accurate to say that no studies detected a difference between groups of children that were exposed to thimerosal vs. those who were not. One large study on 1,047 children from 7 to 10 years of age did find some differences in overall neuropsychological defects (autism in particular was not studied), but the differences were small and oscillated in both directions; of 42 neuropsychological functions measured, more thimerosal exposure was associated with better function in some measures and worse function in others – a small but statistically significant association with tics was observed.[18] Moreover, it is disingenuous and incorrect to say there is no evidence of a link; indeed, some studies have found an association between vaccines containing thimerosal and autism.[19] This is a good example of the real-world messiness of observation and research, in which one must weigh a preponderance of evidence. Are the results suggesting a causal association "real" or are they the result of bias and/or chance occurrence, and what methods can science bring to bear on the issue to distinguish between the two?

Out of appropriate diligence, the issue has continued to be studied by numerous groups. A 2012 Cochrane report on the topic

[16] Stehr-Green P, Tull P, Stellfeld M, Mortenson PB, Simpson D. 2003. "Autism and Thimerosal-Containing Vaccines: Lack of Consistent Evidence for an Association." *American Journal of Preventive Medicine* 25(2): 101–6.

[17] Stehr-Green et al. 2003.

[18] Barile JP, Kuperminc GP, Weintraub ES, Mink JW, Thompson WW. 2012. "Thimerosal Exposure in Early Life and Neuropsychological Outcomes 7–10 Years Later. *Journal of Pediatric Psychology* 37(1): 106–18.

[19] Geier DA, Hooker BS, Kern JK, King PG, Sykes LK, Geier MR. 2013. "A Two-Phase Study Evaluating the Relationship between Thimerosal-containing Vaccine Administration and the Risk for Autism Spectrum Disorder Diagnosis in the United States." *Translational Neurodegeneration* 2(1): 25.

found no association between MMR vaccine and autism after a review of multiple case series, including millions of children,[20] and current consensus opinion is that the tremendous weight of evidence showing no association between vaccination and autism far outweighs the small amount of information that does.[21] The information that does show an association is of poor quality, meaning that due to small size and study design that allows bias, it suffers a likelihood of making type I errors (e.g., detecting differences that are present by chance alone and do not represent a real association) or pick up a confounding association, but not because of an actual causal relationship.

Importantly, Dr. Wakefield's paper was eventually retracted under the accusation of scientific misconduct and fraudulent behavior. The accusation was not that the paper observed a suspicious effect that turned out to be not reproducible – this is a normal part of science, but rather that the authors had purposefully misled its audience, in effect, knowingly propagating a lie. Dr. Wakefield has vehemently denied this accusation. It remains unclear what exactly happened. Whether Dr. Wakefield behaved badly or fraudulently is certainly relevant to his career and integrity, but the ultimate impact on the scientific process is diminished precisely through ongoing study and experimentation in the self-correcting and iterative process of science. Again, this is why the ability to test something over and over is essential to science, and this is how science differs from historical observations or anecdotal experience. Repeated observations and ongoing studies specifically designed to compensate for known sources of error over time were able to generate a preponderance of evidence indicating a lack of causal association between vaccination and autism – contrary to Dr. Wakefield's report.

[20] Demicheli V, Rivetti A, Debalini M, Di Pietrantonj C. 2012. "Using the Combined Vaccine for Protection of Children against Measles, Mumps and Rubella." www.cochrane.org/CD004407/ARI_using-combined-vaccine-protection-children-against-measles-mumps-and-rubella

[21] Centers for Disease Control and Prevention. 2015. "Vaccines Do Not Cause Autism." www.cdc.gov/vaccinesafety/concerns/autism.html

In light of this evidence, one might think that the concerns around vaccination, thimerosal, and autism would have been abandoned. Certainly those concerns were abandoned by the scientific and medical communities (although one cannot speak for every individual); however, such was very clearly not the case for many segments of society. In his presidential campaign, Donald Trump suggested a link between vaccinations and autism in an inflammatory speech. Indeed, President Trump subsequently appeared to be eager to have his administration pursue this question further.[22] This is partially in response to some parents and even autism advocacy groups that continue to raise this concern. This is not an attempt to vilify such groups. On the contrary, keeping the issue alive, as described previously, is in keeping with good scientific practice; indeed, since it has been argued that science is iterative and self-correcting over time, and that there is never definitive proof to a logical certainty, then one cannot absolutely conclude that there is no link between vaccines and autism. However, as explained previously, the association would occur regardless of whether there is a causal link or not, and after taking maneuvers specifically meant to compensate for common sources of error and confounders, the scientific evidence strongly points to no link.

Sticking to the conclusion from initial observation, even in the face of large quantities of high-quality data to the contrary, may actually costs lives. Not only does lack of vaccination result in great harm from infectious disease, but presumably, there are actual causes of autism that we are less likely to discover if our focus and energies are spent on a factor that is not involved.

WHEN THE TRUTH IS TOO OBVIOUS TO TEST

In some cases, for issues of both resources and ethics, science stops short of the very standards that science heralds as essential. Making

---

[22] Specter M. Jan. 11, 2017. "Trump's Dangerous Support for Conspiracies about Autism and Vaccines." *The New Yorker*. www.newyorker.com/tech/elements/trumps-dangerous-support-for-conspiracies-about-autism-and-vaccines

observations costs resources and limited resources often forces a prioritization of what will be investigated. However, there are also issues of ethics. As an example, no one has ever reported a randomized controlled trial of the effects of smoking on rates of lung cancer. Essentially all such studies have either compared rates of lung cancer in those who have smoked to those who do not smoke, or have compared rates of smoking in those who have lung cancer with those who do not have lung cancer. Many such studies have attempted to control for confounding variables, such as making sure that the people in the two groups were of similar age, gender, ethnic background, etc. Nevertheless, no randomized trial has ever been reported, and thus potential confounders may be present at a high rate for a scientific conclusion of such weight.

What would a randomized trial of smoking and lung cancer look like, and what are the dangers of not doing one? Consider recruiting a population of volunteers, randomizing them to two different groups, and then instructing one group to smoke four packs of cigarettes a day for 30 years and the other group not smoke at all (or even be around people who smoked). Assume the population was properly randomized for all other variables (an impossible task, but assume it's done as best as it can be). If lung cancer rates were higher in the group that smoked vs. the group that didn't, it would provide more solid data that smoking, as an independent variable, was associated with (and likely causal to) lung cancer. But do scientists really have to do this? Isn't the data collected thus far overwhelming? Doesn't the FDA now require that tobacco manufacturers label their products as dangerous, and hasn't even the tobacco industry finally accepted and taken ownership of this problem?

It seems exceedingly likely, almost beyond any reasonable doubt at all, that smoking leads to lung cancer. However, in the strictest sense of the word, one cannot rule out alternate scenarios. There is substantial evidence that addictive behaviors are at least somewhat genetically determined. So it is possible that there are genetic elements that both promote cravings for nicotine and also predispose to lung cancer (even without tobacco exposure). It is important to remember that different genes may not be independent variables when it comes to inheritance. If two genes are on the same chromosome and are very close to each other, then they tend to get inherited together. Thus, a gene responsible for a tendency toward tobacco addiction may be next to a gene that predisposes one to lung cancer. In these cases, any retrospective analysis of individuals would find that there was indeed a strong association between smoking and lung cancer, but not because smoking caused lung cancer. Rather, two independent events (one that caused cancer and one that caused smoking tendencies) were associated with each other. To directly test this, one would have to gather a large group of people, all of whom wanted to smoke, and prevent half of them from doing so. If the nonsmokers got lung cancer at the same rate as those who smoked, it would indicate such a scenario – something associated with the desire to smoke (and not smoking itself) would be causing lung cancer.

Is the lack of a randomized controlled trial for smoking and lung cancer really a concern? The evidence of smoking causing lung cancer is not just the immense associative data (albeit retrospective), but also basic biology and biochemistry that show mechanistically how chemicals in tobacco smoke damage lung cells and cause DNA mutations that can lead to cancer. In other words, there is a well-developed web of belief that provides strong support for the conclusion. However, in the strictest sense, the most definitive type of study in humans has not been done. As far-fetched as the objection may seem, the conviction that HRT in women decreased heart attacks was extremely strong, tied into the web of belief, and had much basic research supporting it, before the human prospective trials showed the opposite.

As a specific missive to the reader, I am not suggesting that smoking does not cause lung cancer (in addition to heart attacks, strokes, emphysema, chronic obstructive pulmonary disease, and a host of other problems). Likewise, I am not suggesting that a prospective randomized controlled study be done in this case. Given the data, it cannot be ethically justified. Unlike the HRT case, there is little evidence that smoking has any potential health benefit, and thus the ethics of such a trial would be even more challenging, as there is little to be gained and much potential harm to any participants.[23] Even if one found out that smoking was less harmful than we thought, there is no particular reason to smoke and many reasons not to. However, it does remain true that we cannot ethically apply the most stringent scientific methods to rule out that confounders are confusing us with regards to the causal association between smoking and lung cancer.

So, is the truth too obvious in this case to test? A favorite example, used by biostatisticians, is that there has never been a controlled trial to see if jumping out of an airplane with a parachute gives you a greater chance of surviving than jumping without one. One would need to have two randomized populations and have half jump out of planes with parachutes and the other half without. In this case, as with smoking, the cost of the trial in lives is too high, and the truth seems too obvious to need testing. Still, from the strictest point of view, confounder bias cannot be ruled out without running the trial.

One form of genius in those who perform science in human populations is to find naturally occurring conditions in which one can test such issues in an ethical way. However, as described previously, sometimes this is simply not possible based on decisions we make as a society as to what is permissible behavior of the scientific establishment. This is why it is so profoundly disingenuous when groups like the tobacco industry answer claims that their product costs lives with the response "Well, we don't know for sure" or "It hasn't been proven yet." They object to any law regulating their product until definitive proof is obtained. First, this represents the

---

[23] In all fairness, it must be acknowledged that smoking can be associated with symptomatic relief in some cases, e.g., for ulcerative colitis.

danger of not understanding the limits of science, as proof to a logical certainty is not possible (by science or any other means) due to the issues of holistic HD coherence and underdeterminism presented in the first section of this book. However, it is precisely our societal ethics that prevent us from carrying out the more definitive trial (e.g., RCT), which is the closest we could ever get to the highest standards of scientific proof. Thus, our ethical standards have been twisted to perpetuate the distribution of a product that most likely costs the lives of those who consume it, even when consumed properly. The argument is that we should not ban, or even need to label, cigarettes as dangerous until an RCT is carried out to test the association of tobacco with lethal diseases – but no such trial will ever be ethically permissible to our society. In cases such as these, we must go with the best evidence we have and understand the nature of such evidence and the scientific process that gives rise to it.

PART III

9 Remedies That Science Uses to Compensate for How Humans Tend to Make Errors

Every time you receive what some call a coincidence or an answered prayer, it's a direct and personal message of reassurance from God to you – what I call a *godwink.*

– Squire Rushnell, *When God Winks: How the Power of Coincidence Guides Your Life*

There is a whole series of books, including *New York Times* bestsellers, about how the little "coincidences" we experience in life do not occur by chance – they are actually God speaking directly to us and are called "godwinks." After all, what other likely explanation could there be? One coincidence might happen by freak chance, but so many people have so many stories that seem so unlikely that this must reflect a greater thing, a greater force – this must be the voice of God speaking to us personally. More than 1 million copies of Squire Rushnell's "godwinks" books have been sold, so clearly this idea appeals widely to people. Of course, I cannot rule out, nor can anyone else, that God *is* actually speaking to us by using coincidence as his language – maybe this is just the way that God communicates with humans. Indeed, such is the basis for a vast number of belief systems, the number of adherents to which exceed the number of professional scientists in the world by far (it's not even close). Can it be possible that so many people are wrong?

One must also consider all those things that happen that are not a coincidence. Remember the time you were thinking of a friend and then just a few minutes later you got a text message from that very friend. That was so strange, almost spooky, and it made you wonder about the explanation. Perhaps you and your friend have been so close for so long that there is some communication going on between your thoughts, or maybe you had an actual moment of clairvoyance. However, what you *haven't* noticed is all those times that you were thinking of someone and you did not get a text from them, or you got a text but it was from someone else, or you weren't thinking of anyone and you got a text. Once you take into account all the things that happened that you don't notice, coincidence becomes much less impressive. While one can never rule out that an eerie coincidence is the result of God winking at you, one can say that such coincidences would still happen even if God didn't exist.

Modern science recognizes the fact that humans are prone to certain errors, many of which are described in earlier sections:

- Our poor ability to recognize what random systems look like
- Our poor ability to see probabilities in the context of all the background information (base rate)
- Our inclinations to give meaning to chance occurrences and reflex errors of thinking (e.g., heuristics)
- Our formation of generalizations based on very little data
- Our tendency to seek out confirming information and ignore that which rejects our ideas (confirmation bias and special pleading[1])

As scientists have gained increasing understanding of such problems, scientific methods have progressively evolved to try to mitigate (if not eliminate) such errors. In stark contrast, other belief systems (e.g., godwinks) not only ignore strategies to decrease potential misinterpretation, but embrace and encourage the very situation known to be associated with human error and suggest that this is precisely where we should look for explanations and meaning. This is a fundamental difference in the seeking of coherence – in the rules for modifying the web of belief.

---

[1] Special pleading is one form of human fallacy where a prediction is made but doesn't come true; one then explains away the failed prediction through some special circumstance. Related fallacies include cherry picking data, the "No True Scotsman" fallacy, and moving the goalposts.

The question is not whether unlikely events happen, the question is whether they require an explanation when they do. Do unlikely coincidences have any deeper meaning or reflect an underlying association or cause. If events are occurring at random, i.e., by chance alone, there is no reason to assign any meaning to them at all. In fact, there is every reason *not* to assign meaning to them, as doing so will result in erroneous and potentially damaging connections in the web of belief. A major way to assess this is to get an estimation of how likely things are to happen entirely by chance alone and then to calculate whether the observed frequency is different than this rate. If it is not, then this reflects exactly what the world would look like even if there was no association or importance to the occurrences being studied. However, if occurrences are happening at a higher-than-expected rate, then either our estimate of how frequently chance events occur is off or there is something else going on that requires deeper consideration. Whereas scientific methodologies make the effort to consider rare events in the context of all events taking place, and thus to not waste time focusing on what is just random noise, other systems of belief specifically focus on the chance events (e.g., godwinks).

In 1994, a paper entitled "Equidistant Letter Sequences in the Book of Genesis" appeared in the journal *Statistical Science*.[2] Doron Witztum, Eliyahu Rips, and Yoav Rosenberg presented the argument that advanced pattern recognition approaches uncovered a hidden meaning in a biblical text. It is entirely true that when the text of the Bible was subjected to powerful mathematical algorithms to look for patterns that stood out from the background noise, some incredible patterns emerged that are hard to ignore. The reader should note that *Statistical Science* is a serious journal that undergoes vigorous peer review. Indeed, the authors of this paper made the claim that the patterns in the Bible occurred at a higher rate than would happen by

---

[2] Witztum D, Rips E, Rosenberg Y. 1994. "Equidistant Letter Sequences in the Book of Genesis." *Statistical Science* 9(3): 429–38.

chance alone, indicating a deeper hidden code. So what does this mean? Is it true that they have uncovered a secret code hidden within the text of the Bible, through which god is attempting to communicate with us?

Many systems of human observation would embrace the uncovered code, as many people have. However, scientific methods have evolved specifically to compensate for the human tendency to mistakenly find associations that are not there. Advanced statistical methods have been defined to assess if one is simply falling into a base rate–type fallacy. Regrettably, this is exactly what happened in the case of the Bible code, which was simply an error of neglecting to consider all the combinations of letters that had no message or meaning at all.[3] Brendan McKay went on to show that applying the same approach that was used to find the Bible code to Herman Melville's *Moby Dick* revealed predictions of the assassinations of Dr. Martin Luther King, Jr., Prime Minister Indira Gandhi, Presidents Abraham Lincoln and John F. Kennedy, and Yitzhak Rabin.[4] If one focuses only on the "hits" and ignores all of the "misses," these patterns seem extremely unlikely to have happened by chance, which is what makes them so potent. However, once one examines the denominator of the fraction and the sheer number of possible combinations, then the odds of these patterns appearing becomes very high (almost inevitable), and it removes any reason to assign any deeper meaning to their appearance. All that has happened is that a one in a million event has been observed after looking at one million things.

Scientific practice requires scrutinizing observations by methods known to decrease the rate of error. In contrast, those seeking godwinks and the like, specifically favor approaches that increase the rate of error. The book, *The Bible Code*, was a bestseller, in contrast, the scientific debate and follow-up that discredited *The*

---

[3] Bar-Hillel M, Bar-Natan D, Kalai G, McKay B. 1999. "Solving the Bible Code Puzzle." *Statistical Science* **14**: 150–73.

[4] McKay B. 1997. "Assassinations Foretold in Moby Dick!" http://users.cecs.anu.edu.au/~bdm/dilugim/moby.html

Bible Code is not well known and remains obscure. And despite this strong refutation, the sequel to *The Bible Code* was also a bestseller. To this author's amazement, some have interpreted Brendan McKay's analysis to indicate that real messages are actually to be found in *Moby Dick* (in addition to the Bible), as opposed to a demonstration of the folly of the overall approach. Well, at least that is maintaining coherence through the evoking of auxiliary hypotheses (e.g., Herman Melville can predict the future). However, it still doesn't stand up to scrutiny, and it ignores known sources of error in favor of the fantastic over the rational.

MISTAKEN CORRELATION OF UNRELATED THINGS DUE TO MULTIPLE SAMPLING ERROR

If things truly are correlated, which is to say they occur together at a frequency greater than predicted by chance alone, it is a fair guess that they are related in some way (otherwise their occurrence would not be associated to each other). The errors of perception discussed earlier can result in mistakenly attributing a causal relationship to things that are only correlated but still related in some way. However, if enough things are examined, then correlations will appear even between things that have no actual association. This is related to the base rate neglect fallacy and the lottery fallacy discussed in Chapter 7.

One can find all manner of associations that have strong correlations, that are due entirely to the number of things one measures. For example, in a process called "data dredging," a computer algorithm can randomly compare all manner of available statistics, and some correlations will emerge by chance alone – not because they really have an association with each other, but because so many things have been examined. In *Spurious Correlations*, a comical and highly illustrative book on this topic, Tyler Vigen reported uncanny correlations between annual deaths by bed sheets and cheese consumed by Americans, between margarine consumption and the divorce rate in Maine, between customer satisfaction with Taco Bell and

international oil production, and between use of genetically engineered soybeans and email spam, among a great many other spurious
correlations.[5]

COMPENSATING FOR BIASES AND CONFOUNDERS

Imagine that you have been suffering a bout of insomnia, which is
really starting to have a negative impact on your life. Upon hearing of
your problem, a friend tells you that he also suffers from sleeping
problems and drinks chamomile tea to help get to sleep. You don't
think much of it, but after a few sleepless nights in a row you get
desperate and start drinking tea before bed. Over the next week you
start to sleep a bit better, and so you consider chamomile tea to be a
good sleep aid and recommend it to your friends. You might even
share your experience on a blog and thereby influence many other
people to try the tea. But how confident can you be that it really
helped? Did the tea really cause you to sleep better, or are you just
observing the effect of a regression to the mean (as described in
Chapter 7)? In other words, extreme exacerbations of insomnia will
occur, and you will be more motivated to try new therapies when
these bouts are at their worst. However, even untreated, your sleep
problems will fluctuate over time. Since you are most motivated to
try new things when the problem is at its worst, it is inevitable that
you will observe improving symptoms after trying a new remedy. You
have no way of knowing if the tea really helped or hurt. You could
stop drinking the tea and see if the insomnia returns, but regardless of
what occurs, you can't tell if it is due to other factors.

Sharing our life experiences and learning from the experiences
of other individuals is how we typically navigate much of life. Such
experiences are called "anecdotal evidence" because they are based on
limited experiences that we tell to others or are told by others.
Because each person can only have one set of experiences and only
goes through life a single time, essentially all personal experience is

---

[5] Vigen T. 2015. *Spurious Correlations*. New York: Hachette Books.

anecdotal by nature. Individuals testify before Congress to share their particular life stories and experiences. Parishioners give witness in church to share their personal religious experiences with others. The Internet is saturated with the stories and experiences of individual people. Facebook is likely the single largest exchange of anecdotal evidence ever conceived or created. We not only hear stories of what happens to our acquaintances and families; our daily news is inundated with stories of what happens to other people. We read of a person who died in a plane crash, and so we choose to drive 12 hours to our vacation condo instead of flying.[6]

The limitation to one-time occurrences is by no means only part of an individual's personal experience. In 1929, the New York stock market crashed and the Great Depression soon ensued. In 1932, on a wave of social frustration, the Democratic Party was elected into a large majority. A number of policies were put into place by President Roosevelt and the Congress in the form of the New Deal. The American economy ultimately emerged from the Depression and recovered its vibrancy. Many economists have argued that it was the genius of the New Deal that rescued the American economy; after all, it did indeed recover. Still others have argued for a regression to the mean problem; in other words, no Depression has lasted forever, and therefore, no matter what Roosevelt did the economy would ultimately

---

[6] This is an error of the availability heuristic, because flying is much safer than driving, but the news makes the very rare plane crashes highly "available," whereas the very common deaths by car crash are not reported in the same way. The availability heuristic extends to our memory, in which case it is not the media making things more "available" to our thinking, but the way we remember things that are more remarkable or that stand out in our minds. Thus, the availability heuristic is a basic human cognitive error that compounds our tendency to notice "hits" and ignore "misses" – our tendencies to focus on the top of the fraction and ignore the bottom. This may be why humans are so prone to this kind of error. Our mechanisms of biased observation collaborate with our anecdotal social construct and, along with our cognitive bias, of the availability heuristic. For these reasons, going against this kind of thinking is going against our nature and often can feel very wrong. It certainly defies common sense. Purposeful analysis of where humans tend to make errors and taking measures to compensate for these errors is a fundamental part of scientific methodology. It is also one particular way in which science is different from normal human thinking.

recover. Indeed, some have argued that Roosevelt's policies only made things worse and the Depression would have ended sooner had the New Deal not been enacted. Other economists have argued that World War II ended the Depression and not economic policies at all. The regrettable reality of the situation is that all of the theories are equally consistent with the data, and no one can tell the difference between them. The exact same discussion is true with the more recent Great Recession of 2008; many consider President Obama's policies to have helped, whereas others say they hurt. Either way the economy was bound to recover at some point as no recession lasts forever, but the question is whether the recovery happened because of or despite President Obama's policies.

The previous discussion is not to say that you didn't "experience" the chamomile tea helping – that is what you perceived – you started drinking tea and started sleeping better. But how could one test if chamomile tea really helps with sleep or not? A scientific approach would consist of finding 1,000 people who all had insomnia of the same magnitude and randomizing them into two groups of 500. One group would drink chamomile tea and the other group would drink a fake tea. Optimally, their sleep would be measured by some objective criteria (such as brain wave scans), but in many cases the subjects would report subjective information on how well they think they slept. To decrease bias, neither group would know if they got chamomile tea or fake tea.[7] Moreover, the people running the trial would also not know which people got what tea, to avoid subconsciously giving indicators to the subjects or collecting the data in some biased way. The effects on sleep would then be compared. This is not something you could do on your own; but it is standard practice in science.[8]

---

[7] In some cases blinding may not be entirely possible, as some people will already know what chamomile tea tastes like and will be able to guess what group they are in.

[8] It is acknowledged that like all observation-based understanding, we suffer the problems of induction here – we cannot be sure that what is happening now is what happened to you before, nor can we be sure that the group of people being studied reflect your own biology. However, unless we can go back in time, make

What about the Great Depression or the Great Recession? Because no one can go back in time, and neither the Great Depression nor the Great Recession will ever happen again (in the same society, the same circumstances, and in the particulars of 1929 or 2008), we will never know what caused them or what was the effect of the policies that were implemented in their wake.[9] However, this does not mean that one cannot assess the question using a more scientific approach. In the context of HD coherence, one can make predictions and test them by gathering and analyzing historical data. For example, the hypothesis that austerity policies worked in the Great Recession would predict (all other things being equal) that countries that adopted austerity measures would emerge from the recession faster than those that did not. This type of approach is much more susceptible to confounding by other variables, because different countries have much greater variability than do different people and because no randomization is possible. Stipulating that "all other things are equal" is a much greater stretch here. In other words, even if a difference were seen between countries that implemented austerity vs. those who did not, countries that implemented austerity may have had greater overall debt, different forms of government, different trade balances, etc. Any of these factors may itself have caused different outcomes. However, since they were associated with the likelihood of austerity measures, they gave the impression that austerity policies had an effect. Nevertheless, testing historical predictions of a theory in the context of HD coherence is a more rigorous way of assessing an idea than just through noticing associations through anecdotal experience.

TAMING AMBIGUITY BY ESTIMATING ERROR

How does a scientific approach estimate the risk of making an observational error? How much evidence of a phenomenon do we need to

---

1,000 duplicates of you, and give half of you tea and half a placebo, this cannot be entirely fixed.

[9] This is related to the need for reproducibility described in Chapter 5.

feel confident that what we observe is a real phenomenon and not just a chance occurrence? In other words, how do we attempt to never miss a real association without falling for false associations?

As far as we know, there is no way to avoid error entirely. There is a balance here: one can notice essentially all real things that are detectable but also pick up errors (e.g., notice associations that are not actually there), or one can decrease errors but then not detect all the things out there (in other words, you want your mind to be open to experience, but not so open that your brain will fall out of your head).[10] While there is no perfect answer, modern statistical theory has made great progress in defining exactly how likely it is that we are making an error, giving us the ability to modify our observational activities based on our priorities. For example, what is the cost of missing something vs. the cost of thinking something exists when it doesn't, and how do we choose the specifics of our errors to suit our situational needs? The importance of this cannot be exaggerated, for while we remain always uncertain to some extent, we can know how uncertain we are and how likely we are to be making an error. In other words, we can quantify the chance of making different types of error, giving us much greater ability to modify our actions based on reasoned thinking, rather than resorting to guessing based on anecdotal data.

Let us return to the earlier scenario of flipping a coin. Let's say you want to purchase a trick coin that is weighted – so that it comes up heads on every flip. Hopefully this is because you are a magician or enjoy playing pranks on people rather than because you are a con artist, but for whatever reason, this is what you want to purchase. A purveyor at a magic shop provides you with the coin and asks you to pay for it. Before purchasing the coin, you want to know if the coin is weighted to come up heads, as advertised, so you ask the seller if you can do some test flips of the coin before paying for it. For a particular

---

[10] This quote (or phrases similar to it) is often attributed to Carl Sagan, but has also been attributed to other scholars, many of whom predated Dr. Sagan.

number of test flips that you make and for which it comes up heads, how confident can you be that it really is weighted as advertised and that it isn't a normal coin that just happened to get a run of heads by chance alone?

Assume that you are tossing the coin correctly, that you have interpreted the flips correctly, and that the coin is not changing its properties over time. If tossed an infinite number of times, a fair coin will come up heads 50% of the time and tails 50% of the time. If the coin were in fact a fair coin and not weighted, then the joint probability of it coming up heads on two flips is ½ × ½ = ¼ = 0.25 = a 25% chance. Because flipping the coin does not change the coin, each flip is an entirely independent event. It is a common instinct to fall for "the gambler's fallacy," which in this case would be to believe that the coin is more likely to come up tails if we have just had a run of heads. However, such is not the case as each flip is an independent event; the outcome of any future flip is unrelated to previous outcomes.[11] The probability that the coin will come up heads three times in a row is ½ × ½ × ½ = 1/8 = 0.125 = a 12.5% chance. Likewise, the probability that a coin will come up heads four times in a row is ½ × ½ × ½ × ½ = 1/16 = 0.0625 = a 6.25% chance. One step further is the chance that a fair coin will come up heads five times in a row: ½ × ½ × ½ × ½ × ½ = 1/32 = 0.03125 = a 3.125% chance.

Table 9.1 demonstrates how often a genuinely fair coin will give only an outcome of heads for a given number of flips and how frequently one would make the error of concluding it was weighted when it was actually fair. One very important thing to note is that a decrease in error is achieved with each additional flip, although the rate of decrease starts out quite high and then rapidly diminishes. One goes from a 50% error rate to a 25% error rate when going from 1 to 2 flips (a decrease of 25%), but we decrease error much less when going

---

[11] In other situations, the probability of the next incident is changed by the previous incident (called conditional probability). For example, the odds of pulling a club from a deck of cards is changed after you draw each card (assuming you don't put the card back) as the deck has now changed.

Table 9.1 *Quantifying Uncertainty through Probability Determinations*

Number of Flips	Combined Probability	Percent Chance of Making an Error (e.g., Mistakenly Calling a Fair Coin Unfair)
1	1/2	50.0
2	1/4	25.0
3	1/8	12.5
4	1/16	6.25
5	1/32	3.1
6	1/64	1.6
7	1/128	0.78
8	1/256	0.4
9	1/512	0.2
10	1/1024	0.1

from 9 to 10 flips (a decrease of only 0.1%).[12] So, with a given number of flips that are all heads, if you conclude it is actually weighted, how often will you be mistaken?

Whereas each of us might choose a different error rate with which we are personally comfortable, in the world of professional science the answer is currently clear and unequivocal. In the context of the fair coin example, the answer is a 5% chance of error.[13] To

---

[12] It decreases by ½ each time, but because one is taking a ½ of a ½ of a ½, etc., the absolute amount by which error decreases becomes less and less.

[13] The example given indicates the rate of error based on the number of flips and is used to illustrate how rates of error are related to the sample size of data. However,

understand the basis to of the 5% cut off, one has to give consideration to the historical origin of this determination. R. A. Fisher was a famous statistician who developed much of the basis of current thinking regarding statistical analysis of research. Fisher basically put forward the notion that 5% error in our results was good enough for one to accept that an association was real (i.e., that a difference was not observed by chance alone when no real difference was present).

Consider an example in which one was comparing two groups (e.g., patients getting a new test drug vs. patients getting the existing therapy) to see if the new drug had a different outcome than the existing treatment. In this example, the group of patients getting the new drug had a better medical outcome than those getting the old treatment. The basic concern is that the observed difference happened by chance, and that in reality, there was no difference between the drugs (or the new drug gave an even worse outcome). Fisher and his contemporaries came up with methodologies that lead to the calculation of what is called a "*P* value." A *P* value of 0.05 indicates that an observed difference would occur only 5% of the time by chance alone, if there was in fact no such difference (in statistics terminology this is a type I error, or rejecting the null-hypothesis inappropriately). Conversely, a perceived difference would reflect an actual difference 95% of the time.

This concept can be confusing to people. What does it mean to say "a difference was observed by chance alone when there was really no difference?" How can there be no difference if a difference was

there is also the scenario in which you might be trying to figure out how many flips you should perform before buying the coin. The number of flips required to achieve a size of data that can determine a particular error rate is an essential consideration in designing experiments and is part of what is called a power calculation. Given that larger experiments consume more resources, scientists use power calculations to determine how large an experiment needs to be such that the data set that results is large enough to be able to detect a particular rate of error. Also of consideration is that in the example given, we are assuming that if the coin is weighted, it always comes up heads; in reality, a weighted coin would probably come up heads more frequently than tails, but not 100% of the time. This is also one of the factors that would go into the power calculation and other statistical considerations.

observed? The explanation is that *P* values are used for samples of data that reflect a larger population of data. If one is running a trial of a drug on 1,000 patients (500 patients get the drug vs. 500 who get a placebo), then those groups of 500 patients represent a sample of all of the patients who will ever get that disease. The question is how likely is one to see an improvement in the disease in those patients who get the drug vs. those who don't, if in fact the drug had no benefit, but it happened by chance that patients whose disease were going to do better anyway wound up in the group that got the experimental drug. In such a case you observed a difference in a sample, but there is no difference in the overall population from which you are sampling, you just happened to get a nonrepresentative sample.[14]

There are a number of variables that can affect *P* value calculations, including the degree of the difference observed, distribution of the data, the number of patients in each group, and other subtler data characteristics. Accordingly, there are numerous different methods of calculating *P* value, with different assumptions built in. Using the right method for a given set of data is part of the process of correctly quantifying uncertainty. So, according to Fisher, after the fifth flip of the coin with heads occurring on every flip, you should purchase the coin. This is the first flip where the run of heads would occur by chance alone less than 5% of the time (in this case 3.1%).

So why did Fisher choose a *P* value of 0.05? Was there some objective basis for picking this number? Was there a concern about the feasibility of how many things you can observe or how big a difference would be meaningful? The answer is no. Fisher put this forward because he felt it was reasonable, and other statisticians and scientists agreed. Over time, the $P = 0.05$ criterion has become a standard for significance in science, one that is deeply rooted and dogmatically cemented into the fabric of scientific research. To the extent that this rule is adhered to, it does provide an objective

---

[14] In the context of our coin example, the flips you perform are a sample of all the flips that the coin could ever undergo.

character to scientific observations, in that it is not up to the scientist performing the study to determine what makes an observation "significant." Rather, a P value of 0.05 has been predetermined as an acceptable measure of the rate of error.

Despite the objective nature of having a predetermined and inflexible criterion for a finding that is accepted as statistically significant (real) vs. one that is insignificant (not accepted), the P value of 0.05 was nevertheless arrived at by an essentially subjective process: R. A. Fisher put it forward, and others accepted it. What are the practical ramifications of using a P value of 0.05 as the criteria for significance? In short, it means that effects will be perceived up to 5% of the time when in fact there is no effect whatsoever, or, in other words, in up to 1 out of every 20 observed differences there will be no "real" difference. The type of error we have been discussing here involves accepting the presence of an association, which in reality does not exist (as described earlier, a type I error). However, the opposite error can also occur, where a real association exists but is not detected in the data that is collected (often called a type II error). There are a number of methods for calculating the likelihood of having committed a type I or type II error with any given set of data; however, a P value of 0.05 remains the "gold standard" for a type I error.[15]

One distinct strength of having a P value of 0.05 as the gate-keeper of a significant observation is that it provides a level of apparent objectivity to scientific observation. This objectivity occurs

---

[15] It should be noted that, as here, the P value is often explained by scientists as being the chance of making a type I error. Although popular (and in some ways approximate), this is not precisely the correct definition of a P value. In 2016 the American Society of Statistics published a paper addressing this misconception and defining the P value as "the probability under a specified statistical model that a statistical summary of the data (e.g., the sample mean difference between two compared groups) would be equal to or more extreme than its observed value." This is linguistically complex, so we will use the simpler description of the chance of making a type I error in the rest of this discussion, as an approximation that will serve the purpose of our analysis. The interested reader can find a detailed analysis in the referenced report. Wasserstein RL, Lazar NA. 2016. "The ASA's Statement on p-Values: Context, Process, and Purpose." *The American Statistician* **70**(2): 129–33. doi: 10.1080/00031305.2016.1154108

regardless of the subjective origin of a *P* value of 0.05, because it is an accepted, hard criterion to which the scientific community stringently conforms. However, there is also a distinct downside to this, which is the tendency for binary, black/white thinking. Consider a situation in which one is testing a new drug and the group of patients getting the new drug does better than the group getting a placebo (or a currently approved drug). In this particular example, the *P* value for the difference is 0.06, which is typically described as "statistically insignificant." To many scientists and statisticians, this *P* value of 0.06 actually indicates that there is no difference whatsoever between the groups; in other words, no difference was observed and the groups may be considered identical. This type of binary thinking makes it much easier to state the results of studies as "scientific facts" that are of a yes-or-no nature and allows the generation of webs of belief with the appearance of being on solid and nonprobabilistic footing.

The importance of an accepted "hard limit" of legitimacy (a *P* value of 0.05) really cannot be overstated. It prevents the strong human impulse to change the goal after the fact and to admit a finding as significant when it hasn't been established with sufficient rigor (a regrettable and persistent human tendency).[16] However, at the same time, it seems both myopic and ignorant to ignore all findings and not give them any consideration in one's thinking just because they are only correct 94% of the time. In other words, treating an observed difference with a *P* value of 0.06 as if it's the same as there being no difference in the data at all.[17]

---

[16] This is sometimes called the "moving the goal post fallacy," and is a form of special pleading common to human thinking in which one lowers the standard to whatever evidence they have so as to guarantee a positive finding. The opposite fallacy is "raising the bar" – when people are predisposed to reject an idea and keep discounting findings by ever increasing the criteria required to be valid.

[17] In recent decades, a different approach to statistics (called Bayesian Thinking) has gained much popularity in science, in which one attempts to weigh evidence on more of a continuum. However, while this provides some advantages, it causes other problems as well, which are outside the scope of this work. Nonetheless, frequentist statistics and traditional *P* values remain the major approach in sciences.

The practical dangers in the objective 0.05 cutoff for P values are substantial. On the side of type I errors, what this means is that up to 1 out of every 20 drugs we give to patients has no efficacy. In other words, as many as 1 out of every 20 drugs we test for efficacy will be determined to be beneficial, when in fact it is not. So in theory, 5% (1/20) of the drugs you might purchase may have no benefit at all. Some people have accused big pharmaceutical companies of cynically taking advantage of this issue, by using their immense resources to test just 20 random drugs for a given clinical problem, knowing that by chance alone 1 of them will be approved for public use even though it has no efficacy. The company can then sell that drug to a potentially large and lucrative market, depending upon the nature of the disease being treated, and continue do so until a new study is performed that calls into question the initial finding – if such a study is even ever performed. This is an example of taking mathematical advantage of type I errors.

From a less cynical point of view, if one were to run an experiment 20 times, when no effect was present, one would detect an effect with a P value less than 0.05 in one iteration of the experiment. If an individual scientist (or lab) were to then publish that one iteration of the experiment and ignore the other 19, it would essentially be a fraudulent act of professional misconduct. However, if 20 labs each performed the same general study (unbeknownst to each other), the one lab that found a significant difference might publish its findings while the other labs did not. Since readers of the literature would have access only to the one time the effect was "significant," it would have the appearance of substantial forward progress in the generation of knowledge when in fact it would just be a chance effect.[18] This can

---

[18] This is less of a problem in lab-based basic science studies, because scientists are typically required to repeat experiments over and over again to make sure that a significant P value (<0.05) happens over and over again. However, in human studies and trials, both ethics and resources required may compel a situation in which the study is only done once. Regrettably, even in the lab where there is no excuse for not repeating a study, some papers are published around single (or limited) experiments without rigorously testing reproducibility.

happen due to what is called the "publication bias" or "reporting bias." It is well-known among practicing scientists that journals tend to give preference to positive findings over negative findings. Thus, scientists tend not to report when they don't see an effect, and even if they do, journals tend not to publish such submissions. So while it's very useful and protects against big problems intrinsic to normal human observation, the P value of 0.05 can certainly cause problems as well.

Publication and reporting biases can also lead to what I call a "decreased repetition bias." If an initial experiment is carried out and shows a difference with a P value less than 0.05, the experiment will often be repeated several more times to see if it was just a 1/20 fluke that the difference was seen or if it is persistent and reproducible over time. This is just good scientific practice. However, if an initial experiment fails to detect a significant difference, then one tends not to spend more resources repeating the experiment to rule out that a real difference was missed by chance alone. This tendency is motivated (in part) by the knowledge that it is harder to publish negative findings, but is also likely due to the general human psychological bias of being drawn to apparently positive findings. In any case, certain important associations are overlooked as a result, because when they are missed by chance they are not double checked.

There are distinct real-world situations in which the use of P values by scientists has caused real damage. An example can be found in the development of treatment protocols for kidney dialysis. A great deal of retrospective data suggested that extending the length of dialysis resulted in increased life span for kidney patients. A randomized clinical trial to compare longer vs. shorter dialysis treatments showed a difference, as predicted, but with a P value of only 0.06. As explained previously, this was not interpreted as a difference, with a 1 out of 16 chance of having occurred by chance alone; rather, it was interpreted as "no difference." They did observe a difference in their data, i.e., that one value was different from another, but it didn't make the 0.05 cut-off.

Based on this "no difference" between longer and shorter dialysis treatment regimens, guidelines for patient treatment were changed to give shorter treatment schedules. Basically, the decision made could be articulated as follows. The data indicate that longer dialysis treatment increases the benefit to patients; however, there is a 6% chance that this is a mistaken conclusion, and so because we are only 94% sure that such a benefit would be real, this is not high enough for us to justify longer treatment. Mortality rates rose after this change in practice was made. In retrospect, it was admitted that this decision was ultimately incorrect, and that it likely cost a large number of people their health and their lives. What cost these people their lives was a relatively thoughtless adherence to a P value of 0.05, without weighting the more nuanced meaning and the risk/benefit ratio of applying such a standard.[19]

To maintain objectivity, one does have to have some cut-off for what is a sufficiently small rate of error to conclude that an association is "real" and was not observed by chance alone, but there are different costs to pay depending on where one draws the line, ranging from the cut-off being too stringent to too permissive. Of course, all of us would like to have the lowest rate of error that we can in our observations, so why not require a P value of 0.01 or even 0.0001? Regrettably, making observations costs resources, and at times immense resources. As seen in Table 9.1, after a point, increasing the size (and thus cost) of a study reaches a point of diminishing returns with regards to its statistical power. Running a clinical trial on a drug can cost millions of dollars, not to mention having a very real (and not always beneficial) effect on the lives of the participants. Thus, the need to set a level of acceptable error is a practical, albeit regrettable, reality. That having been said, there are some situations where practical adjustment of acceptable statistical levels of error do and must occur.

---

[19] Twardowski ZJ, Misra M. 2013. "Con: Randomized Controlled Trials (RCT) Have Failed in the Study of Dialysis Methods." *Nephrology Dialysis Transplantation* **28** (4): 826–32.

For example, if one is designing a screening test for infection with HIV, then the statistical cut-off for what constitutes a positive test should be set very low; in other words, the test should detect 100% of cases. This will inevitably come at the cost of some false positives. However, the cost of missing real cases of HIV is that infected patients will be missed, die, and may infect others. The cost of false positives is not zero, as it may cause tremendous distress in those who get the test results, which is why counseling is required; this should be followed by a confirmation test that should be designed to minimize false positive results. Why not use the confirmation test to screen in the first place? Because it would give many false negatives and miss some cases (see section on base rate neglect in Chapter 7).

The practice of controlled trials, repetitions, and large sample sizes partially tames uncertainty but never eliminates it. The use of statistics can then quantify the uncertainty that remains. Scientific practice not only knows and acknowledges that it will make mistakes, but due to the type of statistics theory presented earlier, it can estimate how often it will make mistakes and the rate of error it is reasonable to have about particular observations. Thus, the point is not that science "gets things right" as a matter of practice; rather, science has the best understanding of how often it will get things wrong, allowing a more reasonable amount of confidence and/or skepticism about its conclusions.[20] This is another reason why science will never give the certainty presented in many other systems of belief, as it is part of science to focus especially on the uncertainty – to look it in the eye and to quantify it. Science can be pretty certain about how uncertain it is, and that is where it stops (or at least should stop) in the confidence that it puts into its claims.

This approach is not found in many other areas of thinking, which focus on individual pieces of evidence, without evaluating the likelihood that such evidence represents a real association or is just a

---

[20] The word "confidence" has a specific meaning in the case of statistics; here I am using the word as it is normally used in English and not as a term of art in statistics.

random occurrence. For reasons explained in previous chapters, many things can appear to be so unlikely to have happened by chance alone that they must be imbued with meaning, but this is deceptive and plays to human errors in observation – statistical methods and analysis can let us know how likely we are to be correct in our observations and interpretations thereof.

DISTINCTION BETWEEN SCIENTIFIC PRACTICE AND OTHER AREAS WITH REGARDS TO OBSERVATION BIAS

Americans as a group appear to have substantial confidence in science as a source of new technologies, understanding, medical treatments, etc. Indeed, not only do Americans avail themselves of the products of science, but a tremendous amount of tax dollars go to fund scientific research. Americans also have a strong belief in the paranormal. Based on a 2005 Gallup poll, three out of every four Americans believe in the paranormal, with 41% believing in extrasensory perception (ESP), 37% believing houses can be haunted, 32% believing in ghosts, 26% believing in clairvoyance, 25% believing in astrology, 21% believing in communicating mentally with the dead, and 21% believing in witches.[21]

Focusing on where scientific practice and other areas of belief clash can be useful in making a distinction between scientific and nonscientific thought. As previously stated, there is no reason one approach needs to be right and the other wrong, but there are fundamental differences that do seem at odds. One such profound difference, the focus of this chapter, is how the systems handle the issue of mistaking chance occurrences for important observational evidence. The cost of this issue is pretty high. Americans spend tremendous amounts of money on fortune tellers, psychics, tarot card readers, and

---

[21] Moore DW. June 16, 2005. "Three in Four Americans Believe in Paranormal: Little Change from Similar Results in 2001." *Gallup News Service.* www.gallup.com/poll/16915/three-four-americans-believe-paranormal.aspx

other more exotic means of predicting the future. While it may serve as a simple form of entertainment for some, the majority of people who avail themselves of such services likely do so because they want to have information about the future that they can act upon. In other words, they want increased ability to predict and control, which is also the main goal and utility of science. So the motivations are similar, if not identical, in some ways.

Would you want to know whether a drug your doctor was prescribing for you was helping, not helping, or might even be hurting you? It would seem that most people would answer yes to this question, and this is precisely why medical research and scientific trials exist and why the FDA does not allow pharmaceutical companies to make claims without running controlled studies. Again, as we gain an increased understanding of how humans tend to make errors in observation, scientific practice alters its rules to attempt to mitigate such errors. The slings and arrows of base rate neglect, chance occurrences to individual people, and individual observer bias are mitigated by performing randomized controlled trials in which neither the subjects nor the researchers know which group gets what. The randomization itself is studied to assess other differences between the study groups (other than the experimental difference), in order to decrease (or at least understand) the likelihood of unanticipated confounders leading to an erroneous conclusion. In other words, did the randomization achieve a group in which the only difference (of which we know) is the variable being studied? Statistics provide a clear rate of error and indicate how confident we can be in our findings, giving precise estimates of the possibility that any perceived differences are due to chance alone or that a true difference was missed. Moreover, the iterative nature of science is such that current findings continue to be challenged over time, so that even when errors are made they are, ultimately, more likely to be corrected.

As a general principle, advocates for paranormal beliefs do not practice such maneuvers to address known sources of human error. To the contrary, clairvoyants and fortune tellers tend to do just the opposite. They focus on individual experience, make a number of predictions, focus on the correct "hits," and minimize or ignore "misses." Clairvoyants and fortune tellers are frequently guests on talk shows and television specials, where they demonstrate their uncanny ability to read minds and predict future events. A number of popular television shows and movies provide audiences with vague, strange, and difficult to perceive "oddities" that may be abnormal or paranormal, but no manner of rigorous test is ever applied, at least not of the type described previously that is designed to decrease known sources of bias. Often, pseudoscientific approaches are used. Curious instruments are shown to measure effects and some tests are carried out that appear to be experiment-like; however, there is never any understanding of what (if anything) the instruments really measure and the experiment-like activities are not carried out in a way to decrease error or even give outcomes that can be interpreted. Of course, this fits into the incentive structure of the televisions shows, which make more money the longer they can keep presenting such curiosities. To show that there is really nothing there would eliminate any future shows and thus decrease income. In the words of Upton Sinclair, "It is difficult to get a man to understand something, when his salary depends upon his not understanding it."[22]

From 1964 until 2015, the James Randi Educational Foundation offered a cash prize to anyone who could show paranormal abilities under controlled scientific conditions. James Randi was a stage magician and thus it was his trade to purposefully fool people for entertainment. However, he became frustrated that others were using similar techniques to feign paranormal abilities, and he devoted much of his life to debunking paranormal claims. The word "debunking" is not meant as pejorative in this context; rather, James Randi simply applied scientific methodologies to paranormal claims. As part of his activities, he established the Randi Prize that started as $1,000, but

---

[22] Sinclair U. 1994. *I, Candidate for Governor: And How I Got Licked*. Oakland: University of California Press, p. 109.

which eventually grew to $1 million; despite its magnitude, this prize was never claimed by a single human candidate.[23] Somewhat surprisingly, given the number of psychics and professionals who claim paranormal abilities, there were few applicants over the years, although one assumes such individuals could have benefitted from $1 million in extra income.

Typically, an applicant to the Randi Foundation was subjected to a "preliminary test" in which they were asked to make a number of predictions (in the context of whatever paranormal abilities they claimed to have), and then the frequency of their correct predictions was compared to the rate of getting it right by chance alone. To the best of my knowledge, not only did no one ever win the Randi challenge, but no one ever got past the preliminary test. A typical example was described in an article in the *Guardian Magazine*, in which a famous medium, who had appeared on a number of popular television shows, took the preliminary test.[24] In this case, the psychic candidate had to write out a "reading" (a description of the personality and background of the individual) for each of 10 volunteer subjects whom she had never before met. Each volunteer had to wear strange attire and sit facing away from the psychic, so the physical appearance of the volunteer wouldn't influence the readings. Each volunteer was allowed to review all the completed readings and pick the one that most applied to them. If the clairvoyant had any real abilities, one would predict she should read the people correctly at a frequency higher than that which would occur by chance alone. By chance alone, it would be expected that 1 out of 10 subjects would choose the reading made specifically for them, but in this case, not a single one of them made that choice.

---

[23] Ultimately, scientific neuroimaging technology was able to monitor certain brain patterns around pattern recognition, and as such, ultimately won the prize; however, this is not paranormal but is simply the application of well-understood technology.

[24] French C. May 12, 2009. "Scientists Put Psychic's Paranormal Claims to the Test." *The Guardian*. www.theguardian.com/science/2009/may/12/psychic-claims-james-randi-paranormal

The component of scientific practice that was utilized in the Randi tests was to simply deprive the claimed psychic of cues and data – any information about the subject that one could observe with the normal nonclairvoyant senses. In other words, if the claim was that a clairvoyant ability existed, then just isolate clairvoyant input (eliminating other sources of information); thus, as discussed previously, science is attempting to decrease confounders that may give the appearance of an association when there is none. Many cases that James Randi debunked over the years were outright and purposeful trickery; however, it seems likely that others (perhaps the majority) of psychic claims are actually examples of legitimate errors in human pattern recognition. In other words, psychics may believe that they truly do have psychic abilities; however, they are unaware of their subconscious detection of cues that gives them the ability make seemingly amazing predictions, which they then "confirm" using precisely the natural human biases that science is designed to mitigate.

For example, a cold reading is known to be carried out by an eager psychic and a willing subject. The psychic can make guesses based on the subject's age, gender, appearance, and manner. Incorrect guesses are ignored, but correct guesses are confirmed by the subject and then pursued by the psychic, leading to more and more information. However, when deprived of such cues, psychic abilities vanish. James Randi demonstrated this issue in stunning detail in an interview he once held with a psychic medium and her client, both of whom believed she had been able to read the client's mind, based on her ability to gain information that she could not otherwise have known. Both she and the client experienced that she correctly guessed the names of many individuals in his life. He had kept a tape recording of the session, which was provided to James Randi for analysis. Upon analysis, it was revealed that she had actually guessed a great number of names, 37 names in total, including: Allan, Alfred, Alice, Ann, Bill, Charlie, Colin, Connie, David, Derrick, Eileen, Ellen, Florrie, Frank, Fred, George, Jim, Joe, John, Karen, Katherine, Kevin, Lillian, Lisa, Liz, Lynn, Mark, Mary, May,

Michael, Rob, Ron, Shirley, Sidney, Stanley, Sid, and Steve. Of these, nine were identified as "hits" in that they had special significance to the person being read. However, the identities of such people were son, lodger, brother, nephew, adopted nephew, grandfather, fellow worker, cousin, and neighbor's dog.

The psychic dismissed this criticism in pointing out that one can analyze anything, but that her client felt that what he actually got made sense to him. In the context of previous chapters on analyzing massive amounts of data, it should now be clear that if I gave you a list of 37 names and a hit was counted if a name matched anyone you knew in your life (relatives, friends, acquaintances, and pets), then there are going to be many "hits" by random chance alone. How many of these names hit on someone significant in your life?[25] That it was meaningful to the client speaks more to his perceptions than to any clairvoyance. I think we should accept that both the client and the psychic strongly believed a meaningful reading had taken place, but that this belief was due to confirmation bias – noticing things that seem to confirm and ignoring everything else.[26]

What is true, and what is so compelling, is that both the psychic and the client truly experienced clairvoyance. That is to say, they perceived what they thought to be strong evidence of clairvoyance and "felt" that it was meaningful. What they felt is not in dispute, but what is in dispute is whether their feelings were based on misperception or a real underlying phenomenon. The evidence seems to indicate that this is a case of inadvertent bias introduced by errors in human observation, and all manner of people, including medical doctors and professional scientists, are susceptible to such biases. This is precisely why science has evolved a methodological approach to attempt to decrease known sources of human bias and error – a method that James Randi employed.

[25] This example must be taken in context of language and culture where these names are common.

[26] Randi J. n.d. "Psychic Cringe Fails 2 – The Best of James Randi." *YouTube.* www.youtube.com/watch?v=uq5MtA33OHk

Let's go back to the example of the psychic in the article in *The Guardian* who was deprived of cues and for whom psychic abilities disappeared. Importantly, a great deal of effort went into controlling for a number of potential biases in the study design, the subject pool, and the conditions of the Randi trial. More importantly, the psychic candidate was involved in negotiating the conditions, and she herself agreed that the conditions were fair and appropriate. As is not uncommon, sometime after the test the psychic candidate wrote to the Randi Foundation explaining that the design of the trial affected the subjects, making them "not really free to link with Spirit" and thus interfering with the psychic's normal ability. So what has happened in this case?

This kind of claim testing falls fully into the bailiwick of simple testing of the validity of an observation itself. Whether a claim of a phenomenon is true or false does not depend upon understanding the mechanism, having a grasp on theory, or knowing anything about cause. Rather, claims about the phenomenon are simply claims that an observation is correct that a thing or an effect really exists. Seemingly bizarre or unlikely claims are made all the time, both from within the sciences and without. In both cases, however, how the claims are followed up on is different in scientific practice versus nonscientific practice. In the case of the psychic and the Randi Foundation challenge, an individual claim (supported by a great deal of anecdotal evidence) was subjected to a rigorous scientific trial and it failed. The candidate then engaged in what appeared to be a classic special pleading exercise, where she explained that certain elements of the test prevented the phenomena from occuring. People who claim to have psychic abilities often say that "skeptical testing" or even the process of being observed will inhibit their abilities by creating a "negative energy." Their abilities seem to work only when they are not being tested.

Often, as was the case here, the elements of the trial being objected to are required for the trial to remain scientifically valid. In terms of the working model of science that we have described in this work so far, a claim was made, the evidence did not support it, and then an auxiliary hypothesis was introduced to rescue the claim; in particular, that the application of scientific method itself affects the phenomenon. Thus, the effect of the objection is essentially to render the claim untestable by standard and/or acceptable scientific methodologies. This in no way makes the claim invalid; there is no reason why the "negative energy" of a trial might not interfere with an effect. However, what this *does* do is render a claim such that it can't be tested by scientific methodologies and approaches. Such claims are simply not the "stuff of science" and cannot be assessed by it. This fits into our evolving definition of what science is and how it differs from other approaches to explaining the world. Scientific method progressively incorporates maneuvers to mitigate (or eliminate) known sources of error. The more we learn about error, the more maneuvers we introduce to mitigate it. When paranormal claims refuse (or are unable) to allow their study by approaches that decrease known sources of human error, they cannot be science, and this is one basis for a demarcation to be made between science and nonscience.

SELF-DELUSION AND MISOBSERVATION IN THE HARD SCIENCES

It may come as little surprise to some readers that the claims made by individuals who believe in paranormal phenomena wind up to be a series of perceptual or cognitive errors that give the appearance of clairvoyance when none exists, or at the very least where none can be solidly demonstrated. However, it may be less obvious that precisely the same thing happens in the "hard sciences" – even in physics, which is often characterized as the "hardest" of the sciences.

The late 1800s were a time of explosive new discoveries in physics – ultraviolet radiation, X-rays, radioactivity, and electrons were all described by scientists. In 1903, a famous and highly accomplished physicist by the name of Prosper-René Blondlot added to this chain of new knowledge by announcing a novel form of radiation, which he named "N rays." The detection of these rays was performed by perceiving a change in the brightness of sparks in a special detection instrument, which could be documented on photographic plates. Blondlot's observations were rapidly reproduced by other scientists, leading to approximately 300 published reports by over 100 different scientists who detected N rays emanating from almost all substances and even living things. However, a particular clue to the nature of N-rays was found in the observation that they were not emitted by some metals or freshly cut wood. The discovery of N rays was of such importance that other physicists who had observed the same thing claimed the discovery for themselves, and a special commission had to be assembled to determine who got the credit.

Over time, a problem arose in the study of N rays, because some physicists had a very difficult time observing them. This is certainly known to frequently occur in science, as the specifics of experimental conditions or of detection apparatuses can be quite particular. For example, to detect things in natural phenomena you have to carefully adjust your instruments, and if you adjust them incorrectly, you might miss a critical detail in your observations. The differences in how scientists set up their detection apparatuses can lead to great differences in how scientists observe the same thing, even though what they are observing does exist. Having this understanding can be an important source of new knowledge, as an understanding of the particulars required to observe something can give clues to its properties.

In the case of N rays, most of those who could observe and study them were French physicists, whereas those who could not were mostly German or English. In many cases, this would most likely be due to methodological differences and miscommunication. However, other explanations were also offered. In keeping with the pattern of evoking auxiliary hypotheses to regain coherence between theory and observation, some French physicists evoked the auxiliary hypothesis that the eyes of Germans had evolved in such a way that they could not perceive the effects of N rays, whereas the eyes of French people could. In retrospect, it seems likely that nationalism and patriotism were very much at play in this regard, with French physicists protecting and advocating for the scientific prowess of their countrymen, and vice versa. The claim that French eyes can see things that German eyes cannot seems laughable and absurd in retrospect; however, it is not outside the realm of known human biology. There are certainly people who cannot see certain colors, and this can be an inherited trait, so why not some who cannot see N rays?

Despite the difficulty in detecting N rays in some cases, there were nevertheless a great number of scientists who could easily detect them. Photographic plates were produced which clearly showed the presence of N rays, using well-described methods that were used to discover and document other types of radiation. The study and observation of N-rays was not limited to ivory tower physicists; rather, public demonstrations were done, in which "many in the audience ... had perceived the effects very clearly and 'expressed their pleasure with cries of admiration.'"[27] So, what is one to do regarding the issue of what N-rays are and how one can study them?

One particular physicist, Robert W. Wood, was requested by the journal *Nature*[28] to visit Dr. Blondlot's lab to figure out the problem. To be fair, Dr. Wood was known for doubting the phenomenon and probably went to Blondlot's lab with the goal of discrediting N rays, but this is certainly not a bad trait for a scientist. In addition, the journal *Nature* was published by a society based in Britain (and not France), and as such, some of the previously mentioned nationalism may have been at play. In any case, while it is good and appropriate for a scientist to investigate the claims of other scientists, Wood did act in a somewhat backhanded and deceitful way.

---

[27] Gratzer W. 2000. *The Undergrowth of Science.* Oxford: Oxford University Press, p.14

[28] For what it matters, *Nature* is essentially the most prestigious scientific journal in the world. Although "authority" per se is not supposed to be the arbiter of knowledge claims in the sciences, sources of higher authority still carry more weight, although even they are skeptically tested over time. Some papers published in *Nature* have been retracted or shown to be incorrect.

For example, Wood spoke both French and German, in addition to his native English. However, conversation during the whole visit was carried out in German. Wood pretended to have no ability to speak French during his visit, so that his hosts would feel comfortable saying things of confidence to each other in front of him, which they thought he wouldn't understand. Blondlot and his colleagues put on a series of demonstrations of the effects of N-rays while Wood observed. However, Wood was unconvinced by what he was shown, he simply could not perceive the differences in the spark with or without N-rays. His hosts suggested that Wood's eyes were simply not sensitive enough to detect the differences, as had been suggested previously as to why some could detect N-rays whereas others could not. Again, although this seems absurd, there are many settings in which one needs highly trained eyes to determine an effect. A well-trained pathologist can distinguish cancerous from noncancerous cells when looking through a microscope; an untrained person will "see" the same thing, which is to say the same images will hit their retina, but they will not "see" the same thing, in that they cannot recognize what they are looking at.[29]

Wood suggested that, to compensate for his purported deficiency, that he (Wood) would either block the N-rays or unblock them, and the trained physicists in Blondlot's group with their keen ability to observe N-rays could report when the detector got brighter and when it did not. Wood held the blocking screen (actually his hand) steady to block the N-ray emission, but told his hosts he was blocking or unblocking the N-rays. Indeed, his hosts reported detecting N-rays or not over time, assuming he was moving his hand, when in fact it was blocking the entire time. When he actually did move his hand in and out, he reported that "the fluctuations observed when I moved my hand bore no relation whatever to its movements."

---

[29] Even for the best pathologists there are some specimens upon which there is disagreement if it is cancer or not, or the degree of normal vs. abnormal, but in most cases, it is quite clear.

Wood did not allege that Blondlot and his group were purposely lying, nor did he state the opinion that they were perpetrating fraud; rather, he felt that they had fallen prey to confirmation bias.[30] In such cases, when people know what the outcome of an experiment is supposed to be, they will inadvertently see what they are looking for. In addition, they may inadvertently affect the outcome through subtle changes in method, Alternatively, they may selectively interpret results and/or find reasons to exclude experiments that "don't work," whereas they include those that "do work." What about the clear evidence of N-rays that Blondlot and his colleagues had generated using photographic plates; surely these were not a case of bias, as the differences were clear to anyone who gazed on them and the film itself can't be biased. Indeed, this latter issue caused Wood much concern surrounding the argument that he could not perceive N-rays, as the differences in signal on the plates he saw were so large that he couldn't accept that his eye wouldn't pick them up. Wood was worried that knowing what effects were supposed to be observed, the experimenters would inadvertently expose the plates longer or at a more direct angle, and as such, generate the expected results, when in fact, no N-rays were there to be detected.

Wood continued his trickery around this issue, and unbeknownst to his hosts in Nancy, removed the quartz prism from the N-ray apparatus. By this time, Blondlot had greatly refined his apparatus from the time of his initial discovery, and the prism was supposed to focus the N-rays on the detector. As such, the prism was required for the apparatus to work. Thus, one would predict that the subsequent experiments would fail if the necessary prism was removed. However, Blondlot's laboratorians made the very same observations of N-ray effects that they had reported all along. Presumably, this was the final straw for the good Dr. Wood, demonstrating to his mind that the entire N-ray affair was one big fiasco.

---

[30] Wood did not use the term *confirmation bias*, as it had not yet been coined.

Dr. Wood reported in *Nature* that N rays had been a figment of imagination,[31] that those studying N rays were seeing precisely what they had expected to see; in other words, they had fallen prey to observer and confirmation bias.[32] This report did not settle the issue, as there were some who continue to study N-rays for years (including Blondlot); however, it became increasingly appreciated that under situations designed to control for confirmation bias, N-rays could not be detected. By being iterative and self-correcting, with a focus on employing maneuvers to mitigate error, scientific practice had remedied an error and undone confirmation bias.[33]

Scientists and psychics are both human, and both are susceptible to confirmation bias (also called observer-expectance effect or experimenter-expectancy effect). Such bias can come from multiple sources, but in its simplest form it consists of someone noticing when they see what they expect to see and not noticing it when they don't see it. It can also be described as counting the hits and ignoring the misses.[34] There is a subtle distinction between situations like N-rays and what was previously described using the examples of the red panda in Rotterdam and the battle of Los Angeles. Here one is not misperceiving one thing for another; rather, one is misperceiving an association of one thing with another. Observation bias occurs even with entirely accurate observations of the entities of the natural world.

Unlike many other modalities of thinking, science as a field (and most individual scientists as well) makes a concerted and focused effort to compensate sources of error. In particular, while many psychic and paranormal claims have been shown not to hold up to particular scrutiny, it is neither the paranormal claimants nor their

---

[31] Wood RW. 1904. "The N-Rays." *Nature* **70**(1822): 530–1.

[32] Wood did not use the term *confirmation bias* as it had not yet been coined.

[33] Two excellent descriptions of the entire N-ray affair have been published: Gratzer, 2000; and Klotz I. 1980. 'The N-Ray Affair.'" *Scientific American* **242**(5): 168–75.

[34] An absolutely excellent and relatively comprehensive article showing different aspects of confirmation bias and a great document to read is: Nickerson R. 1998. "Confirmation Bias: A Ubiquitous Phenomenon in Many Guises." *Review of General Psychology* **2**(2): 175–220.

colleagues who have shown such to be the case. It is not normal or accepted practice for members of the psychic community to challenge the claims of other psychics. To the contrary, such critical testing is an unforgivable breach of etiquette. In her excellent essay "Bridging the Chasm Between Two Cultures,"[35] referring to accepted norms of New Age Culture, of which she was a major part, Karla McLaren writes

> ...personal attacks are considered an example of emotional imbalance (where your emotions control you), while deep skepticism is considered a form of mental imbalance (where your intellect controls you). Both behaviors are serious cultural no-nos, because both the emotions and the intellect are considered troublesome areas of the psyche that do very little but keep one away from the (supposedly) true and meaningful realm of spirit.

To the contrary, challenging ideas with deep skepticism is not just a cultural norm in the sciences, but not engaging in such activities is a "no-no." I have many scientific colleagues that I consider to be close friends, and what we do at our annual meetings,is go up to the microphone and try to discredit each other's ideas in a public forum. This is not bad manners – this is what we are supposed to do. It is woven into the fabric of modern science to scrutinize both one's own observations and the claims of others by processes specifically designed to compensate for the sources of human observational error of which we are now aware. It is also woven into the fabric of science to reject one's previous ideas when they are shown to be wrong. Admitting previous error may discredit a scientist's particular idea or theory, but it increases the credibility of the scientist – or at least that is the ideal. As quoted in an earlier section, Carl Sagan described this best: "In science it often happens that scientists say, 'You know that's a really good argument; my position is mistaken,' and then they

---

[35] McLaren K. 2004. "Bridging the Chasm between Two Cultures." *Skeptical Inquirer* 28(3): 47–52.

would actually change their minds and you never hear that old view from them again... It doesn't happen as often as it should, because scientists are human and change is sometimes painful. But it happens every day. I cannot recall the last time something like that happened in politics or religion." Of course, scientists can be small, catty, and competitive people who engage in schadenfreude and behaviors that are not admirable. However, the ideal, which in my view many achieve, is lacking these regrettable traits and working together with those with whom we may disagree, in a joint purpose to unravel the mechanics of nature.

Despite ideals, it is unclear to what extent any given scientist, being human, can entirely avoid observation bias and special pleading for his or her own observations. However, given group dynamics and societies of scientists, those who did not discover the observation (and will likely be less enamored of it) will be less susceptible to the trap of these biases. When one group reports a finding, other groups will see if they observe the same thing. In some cases, if the group is a competitor, or even if there is animus between groups, they may be overly skeptical about the rival group's findings. Skeptics of an idea are no less prone to confirmation bias that the phenomenon may *not* exist than are those who believe it does exist. Of course, many individual humans who are not scientists scrutinize ideas and are skeptical of what they are told. However, many (if not most) people generate beliefs based on a single or small number of experiences, then reinforce this in uncontrolled ways to favor these beliefs (whether advocating or opposing an idea). Science calls out this tendency to be biased and makes it a necessary part of good practice to attempt to compensate for such biases from any point of view. The more we understand the sources of and nature of human biases, the more developed scientific methods become to address them.

Science and its methods evolve as we learn more about the potential shortcomings of previous methods. The recognition of observation bias and our need to compensate for it is a relatively new component of modern science, mostly lacking until relatively recently. This can be illustrated, with both comic and frightening effect, by reaching back into antiquity to remember Galen of Pergamon (Aelius Galenus), arguably one of the greatest medical scholars in the West, who lived from around 129–200 CE. Galen was a leading authority, scholar, and researcher in Western medicine, promoting the diagnosis and treatment of disease arguably to a greater degree than any other contemporary or predecessor. In reference to a treatment that Galen felt was effective, he wrote, "All who drink of this treatment recover in a short time, except those whom it does not help, who all die. It is obvious, therefore, that it fails only in incurable cases." An excellent example of observation bias and special pleading, to be sure!

Galen was certainly considered a scientist[36] in his time. The fact that his activities would not count as good science today only reinforces the argument that scientific methodology is a thing that evolves over time to compensate for new sources of error as we learn about them.

A RETURN TO MAINTAINING COHERENCE BY MULTIPLE PATHS

It is a common narrative by those who describe science that it is a body of observations around which theory is built, and modified to conform to the cold and impartial arbiter of the natural world. As mentioned, and in the words of Thomas Huxley: "The great tragedy of science – the slaying of a beautiful hypothesis by an ugly fact." This view is certainly supported by a general understanding of HD coherence, as was described in Chapters 1 through 3. However, the validity of Huxley's statement rests upon "the fact" being correct. In the last several chapters, we have just scratched the surface of how flawed observation can be and thus how fraught with problems is the very act

---

[36] The term *science* did not exist when Galen was alive, and he was not technically a natural philosopher; however, he was practicing the height of correct scholarship at the time in a field that would be considered a science today (i.e., medical research).

of trying to correctly perceive the natural world. For this reason, when observations don't agree with theories, scientists often doubt the observations. Only after observations have stood up to the current level of scientific scrutiny are they accepted, for now, as being likely. In the same way, theories are only held as tentative "truths" – always subject to modification with the advent of new understanding or information. So must observations themselves be flexible, not as a bedrock of science, but as a malleable arm that can be bent to maintain coherence.

So where does this leave us in the exploration of scientific knowledge claims and how they may differ from other sources of information? It would seem that in actual practice one cannot build a solid edifice of knowledge based on flawless observation of the natural world, as no flawless observation is possible. Science is not and cannot be, as the great empiricists might have imagined it, a progression of knowledge that builds up its truth using observation of the natural world as an infallible arbiter of claims. Science certainly does distinguish itself, in that the natural world *is* the arbiter of all scientific claims; however, human observation of the natural world is flawed, and thus so is our knowledge of what the natural world really is.

To the end, modern science must strike a balance – always poised between theory and observation, knowing that both can be tragically flawed or deceptively correct. In the words of Sir Arthur Eddington:

> But are we sure of our observational facts? Scientific men are rather fond of saying pontifically that one ought to be quite sure of one's observational facts before embarking on theory. Fortunately, those who give this advice do not practice what they preach. Observation and theory get on best when they are mixed together, both helping one another in the pursuit of truth. It is a good rule not to put overmuch confidence in a theory until it has been confirmed by observation. I hope I shall not shock the experimental physicists too much if I add that it is also a good rule not to put overmuch

confidence in the observational results that are put forward *until they have been confirmed by theory.*[37]

Even today there are experimental scientists whom Eddington's notion may shock, who believe that observations are absolute. I would argue that such individuals are mistaken and pay insufficient attention to how observations can be incorrectly made or interpreted. Here Eddington is touching on the holistic web of belief, in that new observations that do not conform to current theory may very well be correct, but they tug and pull on strands of the web tied to previous observations and interpretations that lead to current theory, and this must be taken into account. To reiterate, in science, observations and interpretations can never be arbitrarily changed or distorted just because one doesn't like them or what they imply. Nevertheless, the observation and interpretation must always be considered imperfect (as a function of how humans observe and interpret) as one continues to assess different parts of the web of belief.

The irony here, that has been argued in this chapter, is that while science is often described and distinguished as knowledge based on observation, it is actually other nonscientific approaches that hold observation to be much more holy. Using the earlier examples of clairvoyance and psychics, it is easy to see that their fields accept even scant anecdotal observation as almost absolute and do not challenge its validity. Rather, it is the scientists who hang, draw, and quarter observations to see what holds up to the torture of their skepticism and scrutiny. This is a fundamental difference between science and many other approaches to understanding. Most webs of belief have experience as a large component. Science depends upon experience, to be sure, but scrutinizes it, in an ongoing fashion, with great and unrelenting vigor.

---

[37] Eddington A. 1935. *New Pathways in Science: Messenger Lectures 1934.* Cambridge, UK: Cambridge University Press. (Quotation is from the 1959 Ann Arbor edition, p. 211)

10 The Analysis of a Phantom Apparition, or Has Science Really Been Studied Yet?

THE DISILLUSIONED YOUNG SCIENTIST

Young students of science may choose a career in research for a number of different reasons. Some are driven by an intrinsic curiosity about the world and a love of understanding how nature works. For others, the possibility of recognition and esteem are a driving force. Still others have a meticulous nature, and the notion of gaining and maintaining some control over experimental systems appeals to them. As is inevitably the case, some students pursue science because of the expectations of others rather than from their own interests and ambitions. Finally, some go into science because they've been in school their whole lives, haven't given much thought to what the next step should be (other than moving on to the next grade, as they've always done in school), and really can't figure out what else to do. Given the complexities of human behavior, for many, it is a combination of these factors and additional factors not mentioned here.

Regardless of the reason for entering science as a field, many (if not most) students may find themselves disappointed at some point by what they find. In short, research turns out not to be what they expected. I have frequently had this conversation with young students who were trying to reconcile their preconceived notions of what scientific research was and the reality of what they were now encountering. The work they were doing and that of their fellow researchers seemed "messed up." It was chaotic, did not progress logically, and moved in fits and starts, with many false starts at that. When progress was made, it seemed more like serendipity than the result of any kind of rational thought. Often rationalizations were made up after the fact to account for progress. In some cases (often in the most esteemed and famous labs), the faculty mentors had preconceived notions of how experiments should turn out and were very hesitant to accept any outcome other than what they expected. At times, this attitude would even manifest itself with anger; more than one scientific advisor has been heard saying: "Get back into the lab, and do it over again until you get the right answer!" What's going on here? What happened to the logical and progressive scientific program we have been led to believe exists? The narrative just described is based on my personal experience as the director of graduate studies for a basic science program, but it is borne out by specific studies on the topic that demonstrate not only the misconception of science among students, but the difficulty in changing this misconception even with experience in the lab.[1]

Now there are all kinds of research labs, all kinds of mentors, and all kinds of work environments. Certainly some scientists are more rational than others, some move more on instinct than reasoning, and all need to publish and get grants just to keep the research operation moving, let alone maintain their own jobs and careers. However, even in the best of labs, the disillusioned student may find himself or herself confused: "Why is this lab so different from what I thought it would be?" Often the student blames himself or herself or thinks he or she chose a bad lab in which to work. While this can certainly be the case (some labs and scientists are better than others), often the student will find the same situation when switching to a different lab because the problem isn't with a particular group or project. The problem is with the way in which scientists often portray themselves and how science is typically described by outsiders – both are fictions.

The fiction is not presented with the attempt to deceive and, in many cases, those presenting the fiction are unaware that they are

---

[1] Lederman NG. 1992. "Students' and Teachers' Conceptions of the Nature of Science: A Review of the Research." *Journal of Research in Science Teaching* **29**: 331–59. Cartrette D, Melroe-Lehrman B. 2012. "Describing Changes in Undergraduate Students' Preconceptions of Research Activities." *Research in Science Education* **42**: 1073–100.

doing so. However, those practicing science often publish and report their findings in a way that misrepresents their process out of a matter of necessity, as will be explored later. This fiction causes quite a bit of damage. Because people who study science typically look at its products from the outside and guess at the process rather than experience the process itself, this fiction would inevitably mislead those who analyze science. It has the same effect on the public who perceive science and act (and vote) based on that perception. Indeed this regrettable fiction can cause great harm in a variety of ways.

DISCOVERY IS OFTEN A HIGHLY ILLOGICAL PROCESS, WHICH REMAINS HIDDEN FROM THE OUTSIDE OBSERVER

During my graduate training, I attended a departmental seminar being presented by one of my fellow students in which she described her thesis work thus far. As is common in doctoral training programs, she was presenting a "work in progress" talk to the entire department of immunology. The audience included her lab mates, students, fellows from other labs, and the faculty of the department (including those members of her thesis committee). Presentations of this type are famous for inducing great stress and angst in the student who is presenting. It is expected that the audience will ask critical questions during and after the presentation and that the speaker will have to defend his or her work. As if public speaking were not stressful enough, there is ample opportunity to embarrass oneself in front of one's peers and supervisors, as well as to influence the opinions of those who will ultimately decide if you get a PhD for your hard work or if you leave the program empty-handed after years of effort. Presentations of this type can inflict the kind of trauma on students that causes recurring nightmares (but that also supports the economy by generating tens of thousands of dollars in psychotherapy fees).

I specifically recall this presentation because of the following occurrence. Like most scientific presentations, the speaker used the accepted language of research. She stated that "we hypothesized that TNF-alpha would activate transcription of VCAM." (Translated into English, she was hypothesizing that a particular molecule would activate genes that alerted the immune system to a potential problem.) One member of her thesis committee raised his hand and asked, "Why did you hypothesize that TNF-alpha would work in this way?" This was a typical question to determine the student's background thinking. Basically, he was asking for the logical underpinnings of her hypothesis, for the premises from which she had reasoned to this prediction – how it tied into and sprang forth from the existing web of belief. She paused awkwardly, pursed her lips, and then stated clearly, "Well, we already had a bunch of TNF-alpha in our freezer left over from a different study and didn't have much else to do with it." The room immediately erupted in raucous laughter, drowning out the cries of her faculty advisor: "Don't say that!"

Luckily, after the commotion settled down and the blushing speaker regained a normal skin tone, an atmosphere of good humor returned to the room and she continued to present her data (which were quite impressive). The major lesson here is that much of science actually works in this way – that experiments are carried out because they are convenient or available at hand, not because they are necessarily the best studies to test existing theory. Even more common is that someone has an experiment underway and then runs out of a vital component, so they substitute something else in its place. Sometimes when this is done a different outcome is observed, which leads to a new phenomenon to study. Alternatively, one simply makes a mistake and something really unexpected happens. Indeed, some of the most famous scientific discoveries have been made in precisely this way. Luckily, we have some of these stories.

On September 28, 1928, Alexander Fleming (a Scottish scientist) was working with cultures of a *Staphylococcus* bacterium. One of the plates upon which the bacteria were growing was accidently contaminated by a mold. Fleming noticed a zone around the mold where bacteria were not growing. One might have easily tossed the plate away as a failed experiment because it had mold contamination;

for all we know other scientists had done this many times. Indeed, in Fleming's words, "Now, had I been intensely interested in staphylococcal variation and uninterested in antibacterial substances, I would have cast out the plate, possibly with suitable language and carried on my original program. However, it was the other way about – I was much more interested in antibacterial substances than I was in staphylococcal variation, so I subcultured the mold and the *Staphylococcus*, and proceeded to see why the mold colony should have acted as it did."[2] Fleming had recognized the importance of an accidental finding. The mold was *Penicillium*, which produced a substance later named penicillin, leading to a revolution in antibiotics that changed the world. Luckily for the sake of historical understanding, Fleming described the actual circumstances of the accidental finding[3].

In most cases, narratives for the thought processes of discoveries are not described. We have them for particularly famous findings, because historians become interested and inquire after the fact. For example, one of the seminal developments of modern chemistry was structural models of atomic association and, in particular, the circular structure of a chemical called benzene. These advances were pioneered by a German chemist named Friedrich August Kekulé. Both the notion of structured models in general, and benzene in particular, came to Kekulé in dreams; however, this was certainly not described in his scientific publications. He simply stated, "It now seems appropriate for me to publish the fundamental principles of a theory that I have designed quite a while ago on the formation of aromatic substances." With tremendous and well-earned respect for Dr. Kekulé, "designing" a theory sounds a good bit more purposeful than stating: "It came to me in a dream!" Indeed, if one were to

---

[2] Fleming A. 1944. "Penicillin: The Robert Campbell Oration." *Ulster Medical Journal* **13**(2): 95–122.2.

[3] It should be noted that there have been some challenges to the details of the history of the discovery of penicillin and additional insights into background history that are seldom discussed. Arseculeratne SN, Arseculeratne G. 2017. "A Re-appraisal of the Conventional History of Antibiosis and Penicillin." *Mycoses* **60**(5): 343–7.

submit a grant proposal based on a scientific hypothesis that came to one during a dream, well, let's just say that it may not impress a peer-review panel!

The illogical nature of discovery has long been understood, and many philosophers of science have thrown in the towel when it comes to defining any "logic of discovery" as it is carried out by humans.[4] However, there is a systemic problem that leads to the misconception by the lay public (and many scientists themselves) that the process is imbued with logic. This problem is that science is not reported as it is performed. Since most observers of science only see the work product and not the process itself, no meaningful analysis of the process can be carried out if the product does not reflect the process.

The illogical nature of scientific discovery can be illustrated by the following common joke. A guy named George was driving along one day, when he saw his friend Bill on his hands and knees under a streetlamp on the side of the road. George pulled over and asked Bill what he was doing. Bill mentioned he was looking for his lost keys. George asked Bill where he had lost his keys. Bill responded that he had lost them somewhere in the woods while camping. George, a bit surprised, asked Bill why he was then looking for his keys on the side of the road. Bill responded that he was looking for his keys where the light was best.

Bill, like many scientists, was doing what he *could* do, not necessarily what he *should* do. In science, Bill would find something under the lamp of immense importance, but it would almost certainly not be his keys, and would likely be unrelated to where his keys were.

---

[4] Part of the problem is that since discovery utilizes retroduction, and retroduction is always susceptible to affirming the consequent, then discovery is nondeductive in nature and arguably nonlogical. There are those who disagree (e.g., Mill's methods as applied to discovery of causes). Of note, modern computing power combined with large data sets and the ongoing development of artificial intelligence, have given a recent renaissance to the idea that a distinct logic of discovery may be possible in some contexts and confined to what data sets one can acquire.

PUBLISHED SCIENTIFIC WORK DOES NOT ACCURATELY REFLECT HOW THE WORK WAS DONE

The problem in scientific publishing goes much deeper than the reality that discoveries are illogical. Many stipulate that discovery is a mystery of the human mind (e.g., the light bulb turning on in someone's head) and is not logic governed. However, after the new hypothesis is imagined (or dreamed) the logical component of the hypothetico-deductive (HD) processes is supposed to take hold, and ideas are then tested by an orderly method. This certainly can be the case for individual experiments, focused studies, and incremental advances. This may even be the case for some larger studies. However, this is often not the case, especially in basic research that leads to many new innovations. Nevertheless, scientific papers (the currency of new knowledge) are written so as to make maximal sense to the reader and communicate ideas efficiently; such papers neither aspire to, nor do they accomplish, an accurate description of how the research project was carried out.

Scientific research often thrashes about in fits and starts, first lurching in one direction and then recoiling in another – or perhaps more commonly, simultaneously lurching in multiple incompatible directions. After the fact (or along the way), one is left with a number of observations that can be fit together in different ways and in different orders. When it comes time to report findings in a paper, the observations are typically arranged in a logical and linear narrative that gives the impression of a logical research process, but such is not the case. This fact is well known to professional scientists. Harold K. Schilling, a great scholar of science and religion, presented a highly insightful paper on this topic in 1955, entitled, "A Human Enterprise – Science as Lived by its Practitioners Bears but Little Resemblance to Science as Described in Print."[5] Dr. Schilling's main point – that the

---

[5] Shilling HK. 1958. "A Human Enterprise: Science as Lived by its Practitioners Bears but Little Resemblance to Science as Described in Print." *Science* 127(3310): 1324–7.

popular conception of "science" is a stereotype that is unrecognizable to those who practice science – is no less the case today than it was 60 years ago. He goes on to make the point: "[T]he findings of science are usually presented to students and the public as straightforward, logical developments, rather than in such a way as to reveal how they actually evolved – haltingly, circuitously, with many false starts, and often even illogically... [I]t leaves the uninitiated with a thoroughly misleading idea of the process of science." That this is known to professional scientists is the reason that the seminar room exploded in such laughter when the young graduate student described earlier admitted that she was studying TNF-alpha because it happened to be in the freezer – we all laughed knowingly and then recovered our composure and proceeded onward with the common dialogue of our joint fairy tale.

Dr. Peter Medawar, a Nobel Prize recipient in 1960 for his studies of the immune system, wrote an article and gave a BBC address in 1964 entitled, "Is the Scientific Paper Fraudulent?"[6] Medawar was not suggesting that data, information, or the interpretation of scientific findings were anything other than above board and proper. The scientific content of a paper is honest and not fraudulent in any way. Rather, the fraudulent component arrives with the misrepresentation of how science works and how the process is actually carried out. In response to Medawar's point, Lord Brain pointed out that if the goal is to communicate scientific findings then it need not accurately represent scientific process: "[I]t does not seem to me to follow that the structure of the scientific paper, the object of which is to communicate something to the reader, should necessarily correspond to the logical process by which the discovery was made." Nevertheless, this underscores how one might misconceive what scientists do by reading the fruits of their labors, and not an account of the labors themselves.

In 1977, Julius Comroe, Jr. published the book, *Retrospectroscope*, describing how many famous scientific discoveries were

---

[6] Medawar P. 1963. "Is the Scientific Paper a Fraud?" *Listener* **70**: 377–8.

actually made. The original stories of these discoveries were written after the fact, presenting them in the more logical (and less accurate) light we describe earlier. Dr. Comroe makes the amusing point: "Let scientists use their ... manuscripts to tell how they would have done their experiments if they had proceeded with impeccable logic from the first experiment to the last; this will be comforting to their ego. But let them send with the manuscript a sealed envelope that contains an account that 'tells it like it was' – the envelope to be opened upon the author's death or award of the Nobel Prize, when the ego no longer needs comforting.[7]"

As someone who has read many scientific papers and grant applications, and who has attended many scientific seminars, I do agree with Lord Brain. If one were to attempt to describe how one came to a particular idea or why one did certain experiments in a certain order, it would be highly distracting to the findings themselves (at best) and may be unintelligible to the reader (at worst). Moreover, this issue transcends the work one does in one's own lab. Given the number of professional scientists working today and the length of time a particular study takes, it's very common for new information (relevant to a project) to be published by other groups while the project is ongoing. Thus, the very meaning of one's own experiments may change midstream as a result of the reports of others, as the web of belief continually changes around us due to the number of different scientists studying nature from different angles. It would be folly to stick to your original thinking, purposefully ignoring the new information, and present your findings in the absence of what is now known in the field, simply to remain true to some historical narrative. If anything, this would be the antithesis of a scientific enterprise. That having been said, this in no way diminishes the harm that this process inflicts in the form of giving a false impression to outsiders of what science is, how it works, and what should be expected of it. The facile

[7] Comroe JH. 1977. *Retrospectoscope: Insights into Medical Discovery*. Menlo Park, CA: Von Gehr Press.

communication of data and their interpretation requires a misrepresentation of the nuts and bolts of the process behind the scenes. This is necessary for the web of science to be built and for science to progress; however, that does not mean it is without harm.

The problem of improperly describing scientific process extends to textbooks of science and descriptions of science targeted at a lay audience. Historical ideas that are now considered incorrect are typically presented as part of a scientific narrative, where logical scholars carry out "seminal experiments" that categorically reject old ideas with such clarity that the field as a whole changes. In many ways, this could not be further from the truth. It's a simplified, revisionist history, because what actually happened was messy and chaotic. Much of what we view as "seamless scientific history" really came about by accident and thrash. This is damaging not only to the student of science but also the lay public. In his book describing how DNA was discovered, James Watson states his concern that "There remains general ignorance about how science is 'done.'"[8] He then goes on to describe a process that was not only disordered, but that even betrayed honorable scholarship and a sense of "fair play." In a recent reflection on the work of Shilling and others, Howitt and Wilson point out that "Students may confuse the presentation of a logical argument with an accurate representation of what was actually done. This leads to a view of science that is unrealistic and may even be damaging, as it implies that failure, serendipity, and unexpected results are not a normal part of research."[9]

This problem of improperly describing science remains acute and essential. If those who study science itself do so from the standpoint of the products of science, then they study a fictional account of the underlying mechanics of what actually occurs, and in doing so they may forfeit any hope of fully understanding the thing itself. If students tried to live by the ideal, the fairy tale narrative, it could

[8] Watson JD. 1968. *The Double Helix: A Personal Account of the Discovery of the Structure of DNA.* New York: Touchstone.
[9] Howitt S, Wilson A. 2014. "Revisiting Is the Scientific Paper a Fraud?" *Embo Reports* **15**(5): 481–4.

severely impede their scientific progress. If the lay public accepts the incorrect description of science, it may severely alter how they assess scientific knowledge claims, both by putting too much confidence in the logic of scientific discovery (as opposed to after-the-fact analysis) and in their misperceptions of things having "gone wrong" when in fact, it may be just as it should be.

MOVING PAST ANALYZING THE PRODUCTS OF SCIENCE: OBSERVING THE PRACTICE OF SCIENCE ITSELF.

In 1979, a book of great impact was published entitled *Laboratory Life*.[10] In this work, Bruno Latour and Steve Woolgar, both sociologists of science, reported what they had learned by embedding themselves into the day-to-day workings of a research laboratory. For the object of their study, they chose the laboratory of Dr. Roger Guillemin at the Salk Institute. Dr. Guillemin was a world famous scientist at the time, having won the Nobel Prize in medicine in 1977, and was arguably at the peak of his career. This was basically a study of the anthropology of real scientists at work, and the book synthesized a picture of how science was actually done (as opposed to how it is reported), with a particular focus on the sociological effects of scientific groups.

In *Laboratory Life*, a number of particular observations were made, and theories were synthesized regarding how science works. It is necessary to place these studies in the context of the time, in particular the growth and development of fields studying science that encompass an integration of social dynamics in the analysis of science (including general human anthropology and culture, and issues of politics and group dynamics). This field of study is often called science, technology, and society (STS); in some ways it represents an outgrowth of many of the ideas put forward in Thomas Kuhn's *Structure of a Scientific Revolution*, which argued that scientists'

---

[10] Latour B, Woolgar S. 1979. *Laboratory Life: The Construction of Scientific Facts.* Beverly Hills and London: Sage Publications.

observations were highly influenced by their background beliefs and paradigms. STS acknowledges that such beliefs come from and are influenced by society. Like any program of study, STS is influenced by its own agenda and social structure, an amusing irony of which STS scholars are very much aware.

One of the many observations made in *Laboratory Life* was a confirmation that how science is published and reported is very different from how it is done. This work takes the approach of an anthropologist studying a strange tribe and culture, but draws the conclusion that scientific papers are produced (manufactured) almost as a kind of commodity. Latour and Woolgar argue that HD-type thinking is ruled as much, if not more, by the cost that it entails to modify a belief, and that our beliefs consist of those things that would cost too much to modify. They observe that the meaning of things is highly contextual. In a rather unflattering view of science, they conclude that the stringency with which knowledge claims are tested is simply a function of how much work one needs to do to overcome the objections of others (not one's own self-skepticism) to get credit for "progress." They conclude that scientists work opportunistically, going where the resources and credit are to be found, as opposed to where study of natural phenomena may take them.

The authors of *Laboratory Life* were specifically looking for anthropological aspects of scientific practice; that is precisely what they found. This in no way invalidates their claims, but Latour and Woolgar are no less susceptible to observational bias, confirmation bias, and special pleading than anyone else. It is worth noting that these were purely observational studies, and thus, all of the problems with human observation, which science struggles to decrease, were present for the authors of *Laboratory Life*. That having been said, they made some highly provocative, highly informative, and in some cases transformative observations, all of which support the notion that science is not actually done in the way it is reported. While logic and method may be an essential part of science, and are described in scientific reports and papers, the inner workings of the process are anything but clean, methodological, and orderly. In addition to Schilling's observations that science happens haltingly, circuitously, and with many false starts, it is a social enterprise subject to all the forces and factors of human societal dynamics.

THE SHARPSHOOTING FALLACY OF FOCUSING ON THE "GREAT" SCIENTISTS

Even if the problems described in the previous section did not exist (i.e., science was described in exactly the way it was done), an essential question would remain: "Whose scientific work should we examine to know what science is?" It seems logical that if one wants to know what allows science (and scientists) to make large advances in their ability to predict and control, then one should analyze the practitioners who make the biggest advances. When asked why he robbed banks, Willie Sutton is reported to have famously replied, "Because that's where the money is."[11] Much of the study of science has done precisely that, focusing on the great scientists who have made tremendous advances. Karl Popper used Einstein as his exemplar of good science; after all, who could be more of "an Einstein" than the man himself? In his provocative work *Against Method*, Paul Feyerabend uses luminaries such as Galileo as his exemplars, and he basically rejects any definition of science that would classify Galileo as a nonscientist, which seems on the surface to be a reasonable thing to do. The logical positivists as a group were much impressed by hard-core physicists and scientific pioneers.[12] *Laboratory Life* planted anthropologists into the

---

[11] In his autobiography, Sutton denied making this witty remark.

[12] Logical positivism was a philosophy of science that was dominant from the 1930s until the 1960s. It had the ambition of clearly demarcating scientific statements from metaphysical statements (that had no meaning according to logical positivists). To have meaning, a statement had to be empirically verifiable through observation. While logical positivism ultimately failed to achieve its stated goals, it nevertheless had a profound influence on thought regarding science. Of note, the logical positivists themselves destroyed their own theory, by showing, through careful and insightful analysis and debate, that it could not work. This process of skeptical self-analysis and challenging of one's own ideas is a hallmark of science.

laboratory of a Nobel laureate; they did not choose to study "the average scientific lab," but rather, a famous laboratory.

There is a great deal of ironic symbolism to be found in the debate surrounding what science really is and how we can define it. Scientists study the natural world, and philosophers and historians of science study the scientists themselves. Some philosophers of science have spent their time trying to capture what scientists do, while others have focused more on what they think scientists should do. The irony is that those who study science have repeatedly made the same error(s) in studying scientists as they have observed the scientists making while studying nature. Thus, it is worth reflecting on the validity of the methods and arguments of those who study scientists.

To those who focus on describing what science actually is and how it is practiced, the scientists themselves are the correct object of study. Just as someone who wants to understand nature must use natural phenomena as their arbiter of knowledge claims, so must those who study what scientists do use the actions of scientists themselves as the determinant(s). The historical narrative has been based on the assumption that there is something distinct and special about science; otherwise, why would science have made so much technical progress? Scientists must be doing something that others are not, so therefore, by analyzing what scientists do we can codify scientific methods. The "great scientists" who have made profound discoveries have been the focus of most analysis. This makes a certain amount of sense, for if science is distinguished by the progress it makes possible, then it seems reasonable to study those making the most progress. However, focusing on the extremes in any system can lead to substantial bias; just as much (if not more) can be learned by studying the mundane and common, and the failures as well as the successes.

Failed theories that are rejected can be highly useful in ruling out certain propositions and constitute a "corner of thought" that has been thoroughly explored, even if rejected.

The decision to focus on the great luminaries of science as the exemplars of scientific thinking necessitates some scrutiny itself. Is this a valid maneuver? Do those who make the most progress represent those doing the best science? How could it be otherwise? What seems strange about the focus on luminaries is that scientists are a population, and any population of things has a distribution of characteristics. There is an average for each characteristic, and there are extremes on either tail of the curve. If one were trying to understand the general properties of the population (e.g., to learn what science is by figuring out what it is that scientists actually do), then why wouldn't one focus on the middle of the curve where the average scientists are? Why would one try to learn about the average of a population by studying its extremes? One could argue that it isn't the average we are interested in, but rather the process that leads to the greatest progress; we want to know what works the best, and so we focus on those who have made the most progress, the extreme outliers. This may be supportable, but even so, the conclusions that are typically drawn are generalizations regarding "the whole of science." In other words, scholars of science themselves may be suffering from one of the problems of induction, that the individuals they study may not represent the overall population of their analysis, an issue of which they are (or at least should be) all too aware.

Let us accept for a moment that we are not interested in describing what the average scientist does. Focusing on the extreme of what is the best science and juxtaposing it to what is clearly not science (the other extreme) will give us the best starting point to forge a definition of science. Moreover, we want that definition to guide us in defining what scientists should do in order to make the most progress. We want to write a prescription for the method that will lead to the most advancement; this justifies focusing on the great luminaries of science and what they tend to do. But how confident are we that the scientists making the most progress are those with the best scientific methods? On the surface, this seems like it should be the case. But does this run the risk of falling into the base rate fallacy described in Chapter 7?

There are so many economists out there that any given event in the economy that happens in the next quarter is going to be predicted by someone. When unanticipated and extreme events occur, someone has foretold it. That economist inevitably appears on talk shows, is interviewed for periodicals, and publishes books on his or her view of economics. The economist is heralded as a luminary of economic theory – the great genius who predicted what no one else did. However, we ignore the huge number of economists who got it wrong. In other words, someone was going to make a correct prediction by chance alone, and so focusing on "what they are doing differently" may be studying a phantom, because that person may have no increased ability to predict and just got lucky. There are a great number of different scientists making a great number of predictions, the majority of which are incorrect. By focusing on the scientists who have made the most progress (those who got it right), are we doing the same thing as with our analysis of economists? Are we falling into the same trap when we focus on studying the luminaries of science?

It may rub us the wrong way to ask the question, but is it possible that Galileo just got lucky? Is it possible that we remember Galileo because he happened to have guessed correctly? This feels wrong, but our objection is an emotional one, not a logical one. After all, didn't Newton, Einstein, and others make a number of big advances? Didn't they get it right over and over again? Doesn't this mean they were doing something special? Here we return to the observation by Leonard Mlodinow in *The Drunkard's Walk* that by chance alone, at least one economist should make a correct forecast 15 years in a row, getting it right over and over again, even if their actual ability to predict is no better than random guessing.

How much study has been done of "average scientists"? Or of productive labs that grind away at filling in the details of theory, cataloging predictions of nature, and publishing their findings, making meaningful but moderate contributions?

Perhaps we've been looking in the wrong spot. We've been analyzing the great scientists because that's what has been available to us. In many cases, the groups that publish new, innovative, and even revolutionary ideas represent a boundary-pushing entity that goes against the paradigmatic orthodoxy. Then, other groups who further evaluate the claims and conduct subsequent studies provide a *post facto* critical assessment of the innovative claims. The broader scientific community then serves as representatives of the paradigmatic orthodoxy and provides an adjudication of the novel ideas over time. In many cases, the groups who publish revolutionary (or at least highly innovative) ideas are famous groups who consistently publish high profile papers in top journals; however, it is often not these same groups who perform detailed follow-up studies to explore the deductive predictions of their new theories and provide justification (or lack thereof) for modifying the web of belief; rather, the famous group(s) often move on to the next innovative idea and let other "lesser" labs perform the careful follow-up. This is not a bad thing from the standpoint of the overall scientific enterprise, as it results in a good balance of risky innovation and more measured careful evaluation; however, it does result in an observation bias. In particular, those who are credited with the most progress are not those who are carrying out the most representative science; if anything, they are likely among the more biased and reckless thinkers, and quite possibly the least "scientific."

If the great luminaries of science were just lucky, if it was only serendipity, wouldn't we expect other fields to make just as much progress as science? Aren't those who generate spiritual beliefs and metaphysical philosophies highly innovative and creative individuals? Have there not been a great number of such folks giving rise to new ideas far longer than has been done in "science"? Why, then, have they not made technological progress and changed the world? It isn't because they are focusing exclusively on more abstract philosophical notions. All manner of religious and spiritual beliefs attempt to heal the sick, prevent plague and famine, and predict natural disasters; in short, aren't they trying to address the problems of humankind and seek the betterment of our condition through the ability to predict and control nature? If progress is made just by chance, why haven't these other fields made progress like science has?

This question is partially answered once one abandons the notion that the defining attributes of science are novel ideas, seeing through experience, or any logic of discovery; rather, science is defined by how it evaluates claims once they are made. It is the logic of justification and the practice of science around this logic that bears the weight of reasoning. Such logic is a necessary but not sufficient component of science; the process requires a great deal more, including a dedication to understanding the source of human error and the ever consistent effort to mitigate such errors, to attempt to learn sources of bias and overcome them, to manage pattern misrecognition, to use advanced statistics to understand chance and error, and to adjust and advance what the human senses can observe. However, these things may not be found in the practices or laboratories of the great luminaries of science; the luminaries are too busy inventing (or co-opting) novel and iconoclastic views and pushing them on a reluctant field of lesser mortals, the second tier of investigators, the "follow-up" scientists.

After the luminaries have put forth their ideas, after they have published their papers and given their lectures, their scientific peers then engage in the logic of justification activities. Debate occurs around the existing data. Experiments are repeated by others to see if they make the same observations and have the same interpretations. New data are generated through experiments meant to test the deductive predictions of the theory. In this way the theory is tested, undergoes maturation, and, if necessary, modification, leading to further work, consideration, and dialogue. The theory will likely be accepted by some and rejected by others, but over time the theory is vetted for where it works well, what it predicts, and where it fails, and its value is assessed. It is in this process of following up on the visionary idea where the real stuff of science is done – where the real methods occur. It is in this space that one can find the demarcation between that which is and that which is not science.

When the luminaries are wrong or misguided, as they often are, it is the follow-up scientists who clean up the mess. When the luminaries are correct, as they sometimes are, the follow-up scientists test and vet the theory against the natural world and a wider web of belief. If the luminaries were correct, then it is they who are credited for the progress and not those who really assessed the ideas – and it is the luminaries who are typically studied by those wishing to understand science.

Who exactly are the follow-up scientists? They are the well-trained, indoctrinated, entrenched members of the establishment who read the great journals, attend the luminary's lectures, and then go back to their labs and tinker about to determine if the "genius" to which they have been exposed holds up when applied to natural phenomena. They are the corpus of rational progress, each biased in their own way, but in aggregate they constitute a Bayesian body that twists and contorts and wrestles with the ideas generated by the luminaries. They are the long-term arbiters of the scientific program, the jury of paradigms, the great evaluators. This is where those who wish to understand science should focus their attention. The much less visible follow-up scientists, who repeat and expand the experiments of the luminaries, who develop and mature theories, and who ultimately assess which grandiose claims hold up over time.

The term "follow-up" scientist may at first glance have the flavor of a pejorative, a position that one might accept after failing to be a luminary but not a position one might seek from the get-go. In my view, this could not (or at least should not) be further from the case. Whatever methods we cite to define science, it is the follow-up scientists who embody and employ these methods. The luminaries are clearly a necessary part of the process, as science needs iconoclasts to challenge ideas. However, all manner of human activity and all thought modalities have highly creative, boundary-pushing people bringing forward new ideas. This is, therefore, one of the least distinguishing features of science. What many other fields do not have is the corpus of follow-up scientists to develop, scrutinize, and evaluate the claims of luminaries. Of course, particular individuals may be luminaries from time to time and follow-up scientists the rest of the time; it isn't as though one is born to a certain class (although scientific pedigrees can sometimes feel that way).

Misperceptions about what science really is not only distract science from its true objects of study, but also affect how it is supported. Funding agencies stress innovation and paradigm-shifting activities. Groups want to fund the "big breakthrough." Of course, there's nothing wrong with such aspirations; however, supporting great innovation and trailblazing thought without supporting the underlying engine of logical justification (the incremental work of follow-up scientists) does not create a viable system; both working parts are required.[13]

Indeed, broad and all-encompassing generalizations of how scientists think have been made on the basis of limited information about a few famous scientists. Alternatively, definitions of science have been ruled out (by some) if they don't describe the actions of the famous scientists. Broad sociological observations are made regarding how laboratories function anthropologically, based on the study of a single lab guided by a Nobel laureate. Little attempt is made to capture the average workings of the average scientist and the average scientific lab. This may be the best that can be done with the resources at hand, it is difficult and at times impossible to make a broad study. Nevertheless, it remains the case that the problem of inductive inference afflicts the scholars of science as much as the scientists themselves, inducing a generalization of the whole based on an extremely small sample size, and quite possibly, looking under the street lamp on the side of the road for the keys one lost in the woods miles away, because the road is where the light is best.

---

[13] In the interest of full disclosure, the author of this book is best characterized as a follow-up scientist, at least usually, and this may bias his views.

11 The Societal Factor, or How Social Dynamics Affect Science

R. A. Fisher was one of the most influential scientific thinkers of the twentieth century. He was mentioned earlier for his seminal contributions regarding accurate estimates of the likelihood of an error emerging from a given data set (P values, discussed in Chapter 9). Fisher appreciated that the correlation of two variables only indicated that they have some association, but could not demonstrate causality. In the twentieth century, data began to emerge that people who smoked had a higher rate of lung cancer than those who did not smoke, beginning a debate that would rage for close to a century regarding the carcinogenic effects of tobacco. Unlike most who began to develop the view that smoking tobacco probably increased one's risk of cancer, Fisher became convinced that it was in fact the other way around; he essentially argued that cancer caused smoking.[1] This view, which seems curious in retrospect, was quite logical at the time (and remains logically valid).

Fisher focused on data indicating that the tendency to become a smoker might have a genetic basis (because if one identical twin was a smoker, then the other twin was more likely to be a smoker than chance alone would predict). Fisher was also taken by one study that indicated lung cancer rates were not any higher for smokers who inhaled vs. those who did not inhale (in fact, those who inhaled had lower cancer rates). This was the opposite of what would be predicted if smoking caused cancer by the proposed mechanism. Never one to shy away from publicity, Fisher even wrote a very sardonic piece in which he calculated how many lives

---

[1] Stolley PD. 1991. "When Genius Errs: R. A. Fisher and the Lung Cancer Controversy." *American Journal of Epidemiology* **133**(5): 416–25; discussion 426–8.

could be saved if public health officials would only encourage smokers to inhale.[2]

Fisher's hypothesis was that there was a genetic basis for cancer, and that the presence of cancer (before it became detectable) caused underlying angst in people, which they assuaged by seeking out the soothing effects of tobacco smoke. Thus, smoking was certainly "associated" with cancer, but cancer was the cause of smoking and not vice versa. It has been argued that Fisher was biased by his own love of smoking and that he was being paid by tobacco companies; however, it is important to note that Fisher's "alternate hypothesis" is as valid a retroduction and is as deductively consistent with the observed phenomenon that lung cancer rates are higher among smokers, as is the more popular hypothesis that smoking causes cancer. Nevertheless, over time, more and more evidence began to accumulate that was consistent with smoking causing increased rates of cancer instead of the other way around. Moreover, subsequent larger and better designed studies did not confirm the observation of lower rates in those who inhaled vs. those who didn't; since Fisher himself defined the theory of rates of error (that might have occurred in the original study), one would think that he would have been best suited to be skeptical. However, despite what ultimately became a profound amount of evidence linking tobacco smoke to cancer, Fisher never abandoned his position, one that in retrospect seems to have been very much in error.

R. A. Fisher was not unusual in this respect. Scientists do make errors and have the unfortunate human tendency to fall in love with their ideas and stubbornly stick to them, even after they have been shown to be incorrect. Blondlot never gave up on the existence of N-rays even after all evidence of their existence was discredited (see Chapter 9) and continued to study them privately for the remainder of his career. Priestley never believed he had discovered a new element (oxygen) and continued to study dephlogisticated air.

---

[2] An excellent account of Fisher's career, views, and the entire occurrence of the lung cancer controversy (including relevant references) is found in the article by Stolley referenced in Note 1.

One could fill a whole chapter with examples of personal devotion to a particular hypothesis, no matter how much data ultimately comes out against the idea. Whereas individual scientists often fall prey to this problem, social aspects of science can serve as a remedy, which is the subject of this chapter.

THE INDIVIDUAL–SOCIETY INTERFACE IN SCIENTIFIC PRACTICE

In modern times, scientists seldom work alone. Day-to-day observations, thoughts, and experiments may be carried out in isolation. However, professional scientists (at least academic scientists) typically belong to departments of scientists. They are members of professional societies focused on the topic being studied. They submit their findings for publication to peer-reviewed journals, where their work is evaluated by other scientists working in the same or related fields. Scientists typically present their findings to groups of other scientists at lab meetings, seminars in their departments, during invited lectures to other academic institutions, and at national and international meetings and symposia. Indeed, the very act of communicating one's findings, and what one thinks they mean to a broader audience, is an essential component of maintaining one's scientific profession, for science is a world of "publish or perish." The peer-review process is used to determine how finite resources, in the form of research grants, are distributed to academic scientists.[3] While specific observations or particular studies may be carried out by one (or a few) individuals, the broader determination and interpretation of those studies is a function of a scientific society. Thus, while nature may be an essential arbiter of scientific thought, how nature is explored and interpreted is ultimately decided by a large committee of humans with a complex set of rules and dynamics – fundamentally, a social construct.

---

[3] It is not the intention here to suggest that science only occurs in the academy. Certainly, much science is carried out as part of corporate research and in other less-open settings. As such cultures may vary widely, not much detail about them is given here; however, such groups are nevertheless social groups and thus have social components of group communication and dynamics.

One theme that has emerged in this book is a working definition of science that includes an evolution of methodology in an ongoing effort to mitigate known sources of error to which humans are prone. The errors that have been discussed thus far are errors that individuals make. Because modern science is ultimately produced and evaluated in the context of a social construct, understanding individual error is necessary but not sufficient to grasp the workings of science. Rather, one must also give consideration to the collective reasoning of scientific societies. Do scientific societies exacerbate or mitigate problems of individual error? Are an individual's errors corrected by, or amplified, as a result of their filtration through scientific groups? Do societies of scientists introduce additional errors on top of those made by individual scientists? How do societies complicate the issue(s) of authority in science, serving as their own bodies of authority, in addition to the authority individuals may wield? Deriving any metric of science by analyzing individual scientists (in the absence of group dynamics) will result in a misperception of what science is, even if representative individuals are analyzed.

Science is neither an individual activity nor a social activity, but rather a balanced equilibrium between the two. Both exist and both are necessary, and consideration must be given to an integrated view of individual scientists working on their own and as part of larger groups and communities. To the extent that science focuses on mitigation of human error, group dynamics can help to compensate for erroneous tendencies of individual scientists; at the same time, individual scientists can likewise help to compensate for erroneous group tendencies. Both are required, neither is sufficient, and they must be considered together. These ideas have been developed and explored in great depth by Helen Longino in her excellent work Science As Social Knowledge: Values and Objectivity in Scientific Inquiry (Princeton, 1990).

MITIGATION OF HUMAN ERROR BY SCIENTIFIC SOCIETIES

A great number of assertions have been proposed in the attempt to define science, and many of them have been argued against based on the observation that scientists simply don't practice research in a manner consistent with the proposed definition. This process has rejected practices as a defining characteristic of science because specific individual scientists didn't employ the defined practices. However, we get a different answer as to what science is if we take a broader view of how scientific communities function.

As discussed in Chapter 3, hypotheses can never be entirely rejected even in the face of disconfirming data, as they can always be rescued by evoking an auxiliary hypothesis. Nevertheless, due to problems with induction and confirmation, evidence that rejects a hypothesis is likely more meaningful to the progress of knowledge than is evidence that supports a hypothesis. Moreover, certain types of nonscience and pseudoscience phrase their ideas such that observation can only serve to confirm them and the theories are incapable of being rejected or of being tested rigorously. Thus, the notion of seeking out evidence to reject an idea, and ideas being rejectable, has a strong relationship to what science is.

A separate characteristic that has been proposed as an attribute of science is the notion of Bayesian thinking. Bayesian thinking is a process where scientists weigh the relative probabilities of a theory being true as more and more evidence becomes available, rather than the binary true/false thinking to which humans seem predisposed (simple yes/no thinking). This approach necessitates the ability to change one's mind as a matter of degree of belief, as more information becomes available. It's not just the willingness to abandon a view that one held previously, but to simultaneously hold multiple beliefs as possible, with relative likelihoods for each one.

Both the attempt to reject hypotheses and the adoption of Bayesian notions have been rejected by some as meaningful ways of distinguishing science, because this doesn't seem to be what individual scientists tend to do. As we have discussed, many individual scientists often do not seek to reject their theories, but rather stick stubbornly to their preconceived notions just as most people do. R. A. Fisher never abandoned his position that cancer caused smoking. Despite an overwhelming amount of ultimate evidence that N-rays didn't exist, Dr. Blondlot himself never abandoned the idea; rather, he continued to pursue N-ray studies until his death. When Dr. Wakefield's observation that measles vaccinations correlated with autism was shown not to hold up to controlled scrutiny, Dr. Wakefield did not acquiesce to this view but held to his original theory. This tendency is as true for broad paradigms as it is for particular observations. The list goes on and on, and new members are added to the list every decade. This tendency is a human heuristic that has been named "escalation to commitment" and is a common human error both within and outside of science. Basically, once one invests resources (e.g., money, professional credibility, etc.) in an idea or approach, the tendency is to not abandon the strategy as doing so would admit the previous error and ensure the resources were wasted.

The tendency to stubbornly stick to one's opinions is certainly not limited to those who appear to have come down on the wrong side. When Galileo was advocating that the reason it looked like the Sun was going around the Earth is that the Earth was spinning on its axis, he was well aware of the fact that if such was the case, then there should be a 1,000-mile per hour wind at the equator (according to the understanding at the time); however, there was no such wind. Moreover, Galileo's theory predicted that Earth would be at different positions in the summer than in winter. Thus, the angle of the stars should be different at those times (the parallactic shift). When the measurements were taken, no parallactic shift was ever detected, which could be used as strong evidence to reject a Sun-centered solar system. However, Galileo's response was to say that the stars were so far away that such a shift would be immeasurable. At the time, this must have seemed like an awful lot of arm waving to rescue the hypothesis. There was no way to measure how close or far the stars were at that time; accordingly, Galileo's response fell well outside the realm of testability.

The fact of the matter is that Galileo appears to have turned out to be right, whereas Blondlot was incorrect.[4] The former is heralded as one of the greatest scientists of all time. Scholars try to figure out how Galileo did it, while Blondlot is ridiculed for his errors.[5] However, they essentially engaged in the same thought process. They fell in love with their hypothesis and scrutinized data that argued against it, while embracing and giving an easy pass to the ideas that supported it. The fact that Galileo was correct and Blondlot mistaken is incidental and not a distinguishing characteristic. Scientists, like any other human, can fall in love with their ideas and stubbornly stick to them despite, not because of, the evidence. Individuals are typically not Bayesian; they do not take an unbiased view and adjust relative belief as more and more data come in. Rather, they are devoted conceptual monogamists who are hesitant, and in many cases unwilling, to divorce their intellectual spouses.[6]

HOW SCIENTIFIC SOCIETIES BRING ABOUT BAYESIAN THINKING TO ENGAGE IN FALSIFICATION AND MITIGATE ERRORS MADE BY INDIVIDUAL HUMANS

Scientific societies mitigate non-Bayesian thinking by individual scientists and the problem that scientists fall in love with their ideas and will not abandon them even when tremendous amounts of new evidence arrive that is inconsistent with the idea(s). As was astutely pointed out by Thomas Kuhn, new ideas and paradigms are seldom accepted because the old guard becomes convinced they were wrong; rather, a new generation comes along that is raised in the context of a

---

[4] At least according to our current understanding, which may change in the future, as is the nature of science.

[5] Fisher is a curious exception here, as he is unequivocally seen as a great luminary in statistics theory, but for the particular case of cancer causing smoking (which is not a well-known event), he has been an object of ridicule.

[6] It is noted that I am focusing on the great luminaries of science and not "follow-up" scientists, as I am addressing historical arguments that have focused on the luminaries. As pointed out in the previous chapter, this may be an error. However, there is only limited analysis of how follow-up scientists behave, individually or on average, a situation that may be remedied in the future.

new idea, and the old guard ultimately dies off. Because scientific societies extend over time, new thinkers are constantly being introduced to paradigms and are able to have a fresh look at the thinking and the data, less encumbered by the biases held by the previous generation. In this way societies of science, over time, mitigate the often myopic and tragic devotion that individuals (or even groups) can show for an erroneous idea.

As pointed out by Kuhn (and others), an essential part of scientific paradigms and societies is the act of indoctrinating people to the paradigm. The entire process of scientific education in a particular field consists of learning the language of the field, the definitions of scientific entities, the categorization schemes that are used, the axioms and premises, and the field's holistic web of belief. There is no tabula rasa – no blank slate – for the professional scientist; one learns the history of the field through the filter of the present. It is difficult, if not impossible, to begin one's career as an iconoclast. The very act of gaining credibility necessitates some intellectual lip service to the field and the dominant paradigms. Those young visionaries who do make great breakthroughs are typically trained under well-entrenched members of the establishment and often carry out their rebellious studies in such protected environments.

Nevertheless, if a real and reproducible natural phenomenon exists that is inconsistent with our best theories, such nature cannot be erased and can only be ignored for so long. Individuals can render it meaningless either by dismissing it, or more typically, by just looking past it. However, subsequent individuals will rediscover it, over and over, until its impact is ultimately felt. It is in this way that ongoing societies of scientists mitigate the human tendency to become convinced of an idea and then filter data to fit it. Although generational change in scientific societies may seem glacial, it does change in a way that works against individual bias, special pleading, escalation to commitment, etc. This is one of the reasons why science must be a sustained effort over time, and longer than the human life span, to allow for such successions. In the somewhat morbid words of the famous physicist Max Plank, "Science advances one funeral at a time."

As a group dynamic and in a societal context, science ultimately does behave in a Bayesian and Popperian fashion, even if individual scientists do not. Thus, when taken in a social context, rejecting Bayesian and Popperian criteria as a definition for science because individual scientists do not behave in that way is a mistake of analyzing the component parts of science (individual scientists) and not the whole enterprise. It doesn't matter how stubborn Wakefield or Blundlot or anyone else was in their beliefs. It doesn't matter that proponents of theories refuse to change their minds even when mountains of data are inconsistent with the idea they support. What matters is that, over time, as new scientists evaluate the data, scientists who are less programed to a paradigm and less stubbornly invested in a single point of view ultimately challenge theories that don't line up with repeated observations.

At the end of the day, scientific societies serve to compensate for lack of self-critical thinking by the individual. This is precisely what Latour and Woolgar observed in their anthropological study of a working lab described in Chapter 10 (i.e., the book *Laboratory Life*). The criticism that the stringency with which knowledge claims are tested by scientists was simply a function of how much work one needs to do to overcome the objections of reviewers may be evidence of lack of self-critical thinking. However, through the looking glass, it is simply espousing the virtue of scientific societies in forcing critical thinking. The objections of the reviewers that scientists had to do experiments to overcome (or not, depending on how the experiments turned out), are the critical thinking of others within the society forcing rigorous evaluation of the hypothesis on those testing it, even if the scientists testing the hypothesis are not inclined to do so themselves.

PREVENTION OF SCIENTIFIC PROGRESS BY SCIENTIFIC SOCIETIES

As laudable as the ability of scientific societies to overcome individual bias may be, it comes at a cost, and at times, a very high cost.

Because ideas may change over life spans and not minutes, progress can be slow and maddening to those whose lives are spent trying to effect change. As societies are made of individuals, and as individuals have profound biases, so do societies express profound bias.

A remarkable demonstration of such societal behavior is the famous case of puerperal fever (better known as childbed fever) among women delivering children in the mid-1800s. Hospitals across Europe had maternity wards, to which women would be admitted to deliver their babies. Childbed fever was a disease that would afflict some women who had just delivered, resulting in extreme illness. Mortality rates from childbed fever were quite high, in some cases as high as 30%. The causes of childbed fever were unknown, but a young doctor named Ignaz Semmelweis, who worked in a general hospital in Vienna, Austria, took note of some striking characteristics of the disease. First, this disease seemed to be related to hospitals. It was well known that women who delivered outside the hospitals suffered this disease at much lower rates. This was not an issue of socioeconomic status, as even destitute women who delivered in the streets of Vienna still had lower rates of the disease. Another issue that caught Semmelweis's attention was that there were two different maternity wards in the hospital, one of which had a much higher mortality rate from childbed fever than did the other. This fact was so widely known that delivering mothers would beg to be admitted to the unit with the lower mortality. The difference troubled the young Semmelweis, and he launched into a detailed juxtaposition of the differences between the two wards to try and ferret out the reason. As it turns out, the ward with the higher mortally rates was less crowded, so overcrowding didn't seem to be the answer.[7] A second difference was that the

---

[7] It is interesting to note that it would have been logically consistent to conclude that a higher density of humans packed together had a protective effect, that perhaps some people consumed some of the fever from other people. In other words, Semmelweis might have increased the patient density in the more lethal ward to see if this helped; however, this appears not to have been considered, likely because the theory of disease at the time focused on "miasma" (disease caused by foul air) and less crowding and ventilation was considered health promoting and not the opposite.

ward with lower rates of fever was staffed by midwife students, whereas the ward with the higher rates of fever was staffed with medical students. Why this should lead to different rates of fever was unclear.

In 1847, Jakob Kolletschka, a professor of forensic pathology for whom Semmelweis had great admiration, died from a wound he acquired from the poke of a medical student's scalpel in the course of performing an autopsy. When an autopsy was performed on Kolletschka, he had many findings that greatly resembled childbed fever. Semmelweis drew the connection that some material from cadavers was causing childbed fever. The reason that women were getting childbed fever at a greater frequency in the ward staffed by medical students, Semmelweis hypothesized, was that the medical students were transferring the disease from cadavers to pregnant women (autopsies were only carried out by medical students and not midwives). In response to this theory, Semmelweis introduced a policy of medical students washing with chlorinated lime (calcium hypochlorite) between autopsy work and the delivery of children. After this implementation, mortality rates from fever dropped 90% in the ward attended by medical students, and reached the same rate seen in the ward attended by midwives. In the next year, Semmelweis added the washing of instruments that came in contact with patients, and childbed fever essentially disappeared from this particular Austrian hospital. Later, when Semmelweis's students brought this technique to other hospitals, death rates dropped to similar effect there. By any standards, this accomplishment was a profound achievement, one that not only saved a great number of lives and prevented a great deal of suffering, but gave science a fundamental insight into the cause of childbed fever – at least it should have. While the Nobel Prize had not yet been created, one might speculate that Dr. Ignaz Semmelweis would receive great recognition for his accomplishment; however, such was regrettably not the case.

The medical establishment soundly rejected Semmelweis's findings. Several factors were involved in this rejection, and it should be recognized that Semmelweis published his findings haphazardly. He wrote letters to other physicians, and his students published reports, but he himself did not publish any report until 1858, and he did not write his main work until 1861. Nevertheless, his findings were almost uniformly dismissed as insignificant on a number of grounds.

First, Semmelweis's interpretation of his findings flew in the face of almost all dogma regarding disease, which held that disease was caused by an imbalance in human body fluids and was transmitted through the air by foul vapors. Each person's disease was individual to his or her particular imbalance. It would have been hard, if not impossible, to fit Semmelweis's findings into this dogma. Strikingly, a major objection to his findings was that because he had no theory to explain the mechanism by which disease could be transmitted, how could his observations be correct? It is important to remember that germ theory – the idea that illness was spread by microscopic bacteria and viruses – had not been introduced at this time. How could cadaveric material possibly transfer disease, as there was no entity in the material that was capable of inducing illness? This is another example of where the classic and misunderstood description of "scientific method" doesn't conform to what science actually does. Observation is not a perfect arbiter of our knowledge claims, as observation can be in error – indeed, one must challenge observation. Hypothetico-deductive (HD) coherence can be maintained by changing theory or ignoring data, and both are done with some frequency, as explored in detail in earlier chapters. It also seems likely that there was an emotional component to the rejection, as physicians were none too eager to acknowledge that their practices were dirty, that they needed to wash more often, and that they had been killing thousands of patients each year due to ignorance and lack of hygiene.

Semmelweis was also whipsawed by the British medical establishment, which accepted his observation (if not his explanation) as proof of the validity of a previous theory held in Britain that childbed fever was indeed contagious. However, the British thought the fever was due to miasmas coming out of the dissecting room and causing humoral imbalance in new patients. This may have reflected a misunderstanding of Semmelweis's findings as much as a rejection of them. And Semmelweis may have been partially to blame because of the manner in which he published his findings. Nevertheless, on the Continent he was seen as an idiot and in Great Britain as someone who had contributed nothing new because he was ignorant of existing British wisdom.

Semmelweis was dismissed from his post, likely for reasons unrelated to this particular issue but rather because of great political instability in Europe in general, and in Austria in particular. Additionally, Semmelweis may have been mistrusted by Austrian physicians because he was of Hungarian background. Following his dismissal, Semmelweis resorted to writing increasingly vitriolic letters, accusing those who did not embrace his theory of essentially murdering patients. He became emotionally unstable and was institutionalized in an insane asylum, where he died 2 weeks after being admitted at age 47, probably the result of being beaten by guards.[8]

The resistance of science to change is not restricted to those ideas that ultimately are correct. While the scientific community was very slow to come around to Semmelweis's correct understanding, it has also resisted new ideas that later proved to be incorrect and could have been very damaging. In the early years after the AIDS epidemic had been recognized and when HIV was first isolated from AIDS patients, Dr. Peter Duesberg, a well-respected virologist, was highly skeptical that HIV was really the cause. He believed that the toxicity of the drugs used to treat AIDS was actually causing AIDS. He voiced a number of objections to HIV being labeled as the cause, not the least of which was the existence of an immunosuppressive syndrome (very much like AIDS) that afflicts young males and occurs in the absence

[8] The issue of Semmelweis has been analyzed by many and for different purposes. In his work "Philosophy of Natural Science" Carl Hempel used Semmelweis as an example of scientific method. Here Semmelweis is used as an example of social dogmatism in science.

of any HIV infection.[9] To Duesberg, this rejected the entire HIV hypothesis. This was logical and Popperian; indeed (if AIDS occurred in the absence of HIV, then HIV could not be the cause of AIDS – or at least not the sole cause; a single negative can reject a hypothesis more quickly than confirmatory evidence can prove it). The field did not favor this thinking, evoking the auxiliary hypothesis that these young males were suffering a different disease that was "AIDS-like," but not AIDS, and as such, this was not evidence to reject HIV as the causative agent of AIDS. However, Duesberg was unwilling to accept the auxiliary hypothesis that the disease that resembled AIDS was a distinct disease, and thus the lack of HIV in that setting still rejected the HIV hypothesis in his view. Duesberg could also point to people who were infected with HIV but who had not developed AIDS, but he was unwilling to accept the explanation (auxiliary hypothesis) that HIV had a long incubation time from infection to symptoms of disease, or as we subsequently have come to understand, that some people have a genetic resistance to AIDS. Thus, Duesberg put forth scientific arguments against HIV being the cause of AIDS, consistent with the best tenets of scientific reasoning. However, as more and more evidence accumulated, culminating in the incredible efficacy of anti-HIV drugs in preventing AIDS, Dr. Duesberg's objections were less and less credible. Like Fisher in the case of smoking and lung cancer, he focused specifically on the few data points that appeared to reject the main hypothesis. Also like Fisher, instead of accepting this increasing evidence as it accumulated, Dr. Duesberg has stuck to what now seems like an absurd position.[10,11] However, earlier on,

---

[9] Smith D, Neal J, Holmberg S. 1993. "Unexplained Opportunistic Infections and CD4+ T-lymphocytopenia without HIV Infection. An Investigation of Cases in the United States. Centers for Disease Control Idiopathic CD4+ T-lymphocytopenia Task Force." *New England Journal of Medicine* **328**(6): 373–9.

[10] Duesberg PH, Mandrioli D, McCormack A, et al. 2010. "AIDS since 1984: No Evidence for a New, Viral Epidemic – Not Even in Africa." *Italian Journal of Anatomical Embryology* **116**(3): 73–92.

[11] Duesberg PH. 1998. *Inventing the AIDS Virus*. Washington, DC: Regnery Publishing.

his objections were logical and reasonable, yet were dismissed with emotional venom by the establishment, likely prematurely and with questionable arguments.

In the case of Duesberg, the slow dogmatic stubbornness of societies of scientists prevented the field from going down the wrong track that would have delayed (or prevented) the breakthrough life-saving treatments that have now been developed for HIV and AIDS. In the case of Semmelweis, the stubbornness of scientific societies costs the lives of thousands of women and their babies. However, in both cases, the societies serve as a stabilizing factor, with an ultimate Bayesian filter preferring evolution of ideas over revolution of ideas. This can have good and bad effects, but underscores one role of society in science – that ultimately, the society came around to the greatest HD coherence of theory and evidence through an iterative and self-correcting process grounded in observation of the natural world.

SCIENTIFIC SOCIETIES CAN ALSO AMPLIFY ERRORS OF HUMAN THINKING, OR GENERATE BRAND NEW EMERGENT ERRORS ON THEIR OWN

Ironically, while scientific societies can, over time, mitigate individual errors of confirmation bias, special pleading, and escalation of commitment, societal judgment can also amplify such errors in the short term. Scientific societies can simultaneously mitigate and exacerbate the same problem. Even though evolving groups of scientists decrease the effects of individual bias by allowing new minds to reevaluate problems over time, the social structure of science also allows faulty thinking to extend far beyond the personal thinking of a given individual.

The term "groupthink" was first popularized in 1972 by Irving Janis, who analyzed ignominious errors that came out of the group analysis of situations.[12] Of particular note is Janis's assertion that groups of people thinking together can make specific types of errors

---

[12] Janis IL. 1969. *Personality: Dynamics, Development, and Assessment.* New York: Harcourt, Brace & World.

that individuals might not make on their own. According to Janis, groups generate an environment where conformity is forced upon individuals. Ideas that differ from the group consensus are "self-censored," putting a strong damper on alternative interpretations or creative thinking. Lack of clear dissent or even silence is viewed as agreement, leading to an exaggerated feeling among group members that more individuals agree with the group position than is the case. It can be very hard for individuals to be a first voice of opposition against something that everyone else appears to hold as true. Going out on a limb can be a scary act, opening one up to ridicule and marginalization by the group. The group works to rationalize away dissenting views or data that appear to disagree with the group, which has a strong resemblance to special pleading by individuals who pay close attention to observations that support their preconceived notions, but who ignore data that contradict what they already think. A kind of us vs. them mentality can result, and those who disagree with the group position may be judged as less intelligent, ignorant, or even as morally inferior to the group.

Regrettably, modern scientific endeavors can have many characteristics of groupthink behavior. It is a sad dynamic, consistent with stereotype, that some scientists label those who do not embrace the canons of science as being ignorant, stupid, and inferior. Evidence that conflicts with scientific dogma is often cast out as quackery. Should those who present such evidence be nonscientists, they will be written off as either uneducated buffoons (in the best case) or as cynics trying to manipulate a gullible public for notoriety or money.[13] Should those who present such evidence be from within the scientific establishment, their credentials and experience may be challenged;

---

[13] The fact that scientists have this tendency does not discount the situation that there are in fact many people who will purposefully misrepresent knowledge claims to gain power and money. Whether they really believe the claims or whether they simply have a different opinion or are cynical liars, they do profit and can do great harm to others. Of course, since I am a professional scientist, my opinion on this could be attributed to groupthink, and if such were the case, by my own definition, I wouldn't be able to know if that were so.

they may be branded as having lost their way or, in some cases, as having lost their minds. In some cases, their previous and subsequent opinions, even of a noncontroversial nature, are given little consideration, their credibility having been damaged by their antidogmatic views. Scientific societies can be allergic to new ideas, aggressively persecuting those who bring the ideas forth.

The dogmatic behavior of scientific groups occurs just as frequently when the scientific establishment turns out to be right as when it turns out to be wrong. Peter Duesberg has stuck to his view that HIV does not cause AIDS and, in doing so, has become a persona non grata in science. That he appears to be incorrect in his HIV views does not necessarily mean that the greater group of scientists is correct in making him a pariah. This also happens to people who ultimately turn out to be correct. As much as Galileo and Copernicus are credited with discovering that the Earth orbits around the Sun, this idea had been introduced to the ancient Greeks by the astronomer Aristarchus of Samos, who it appears (although evidence is limited) to have been basically dismissed out of hand.[14] Of course, Galileo was forced to recant under threat of torture by the Vatican. Louis Pillemer's seminal discovery of a new way that the immune system fights infection was dismissed so as to discredit him, prevent its publication, and remove his scientific funding, leading to his suicide.[15] Yet ultimately his findings and interpretations were discovered to be correct. Therefore, while an evolving scientific society mitigates individual bias over time and over scientific generations, it may also exacerbate it at a particular time and in some ways through societal dogma and groupthink. It is always acceptable, indeed compulsory, to disagree with people based on the web of belief; however, discounting someone's view out of societal dynamics and peer pressure is anathema to the ideals of science.

[14] This is largely speculative as records from antiquity are incomplete and somewhat unclear.

[15] Lepow IH. 1980. "Presidential Address to American Association of Immunologists in Anaheim, California, April 16, 1980. Louis Pillemer, Properdin, and Scientific Discovery." *Journal of Immunology* 125(2): 471–5.

Janis also explored how groupthink can cause a group to feel it has a heightened sense of morality that is immune from normal rules of behavior; the group believes it has an elite status that deserves special privilege. This dynamic can lead to feelings of invulnerability, resulting in excessive risk taking and reckless, even sociopathic, behavior. Janis was mostly analyzing political groups that have made egregious decisions. It is unclear that these latter characteristics apply to modern science as a general endeavor. However, there have clearly been egregious violations of human rights carried out by groups engaged in scientific endeavors. One must be mindful that, as with all groups, the potential dangers of groupthink go beyond mistaken knowledge considerations.

Although groupthink and its associated theories are relatively modern terms, the appreciation of such effects goes back a long way. When Francis Bacon wrote the *Novum Organum* in 1620 to espouse the virtues of inductive observation, he defined a series of four common sources of error (he referred to them as "idols"). Among these was "idols of the theater," which reflected the regrettable influence that existing dogma, if not groupthink, could exert on the individual mind.

Since Janis's early description, it has been pointed out, and rightly so, that the specifics of groupthink are fairly speculative, with very limited proper studies to test the theory. Nevertheless, some concrete examples persist. Cognitive psychologists have challenged Janis's interpretation of such group phenomena and have presented new theories of why such group dynamics occur (these go beyond the scope of this book); however, it has not been disputed that groups influence individual thinking in profound and varied ways. Given the history of behavior by scientists and their societies, it would be difficult to justify a position that some dynamic of this type is not playing a role in science, and at times, potentially a large role.

THE DANGERS OF SOCIETAL AUTHORITY

In many ways, authority is the unequivocal enemy of science. As has been argued, scientific knowledge claims are ultimately arbitrated by observations of natural phenomena, retroduced hypotheses and theories, auxiliary background assumptions, and the web of belief to which it has given rise. That is not to say that at the end of the day expert opinion of what the data mean is not still at play. However, expert opinion, or authority in the absence of evidence, is a different kind of thing. As such, the simple justification "because I said so" has no place in science. Of course, expert opinion may be all that is left if the testing of an idea is neither technically nor ethically feasible, or if the necessary resources to test it simply cannot be justified. However, authority-based knowledge is never the goal. Many humans appear to have a strong natural gravitation to authoritative figures, be they individuals or groups. Since scientists are humans, this affects science. Sadly, humans find confident statements made by authority figures to be incredibly attractive, if not fundamentally compelling. Ironically, those who admit to at least some sense of fallibility or uncertainty, who are the closest to being credible and realistic, are less compelling than those who speak with unrealistic and even delusional fabrications of certainty.

It has been argued that the genesis of science depended upon the free (and unpunished) questioning of authority and the ability to attack the ideas of one's mentors. To the best of our knowledge, most ancient societies were profoundly hierarchical – typically ruled by the authority of a single monarch. In many cases the monarchs ruled by divine right, if not claiming personal divine status. Authority may have been exercised through an oligarchy and/or a bureaucracy; nevertheless, the authority was absolute. Whatever the authority said was true was so because he or she said so – to disagree with the statements or judgments of the authority would be heretical, and in many cases, punished severely.

It is meaningful to note that the genesis of Western science occurred in a society with less stringent authority, which had greater tolerance of free thought and dissent. Unlike many of the centralized and divinity-ruled empires before it, ancient Greece was a series of loosely affiliated city-states, some of which developed a democratic governmental structure (most notably Athens). Commerce and trade across cultures allowed an influx of different ideas, religions, and views. Debate, dissent, and disagreement were not only tolerated, but in many cases encouraged and even revered.

Anaximander of Miletus was an ancient pre-Socratic thinker who made tremendous contributions to the development of Western thought. It has been argued that one of his most profound gifts was the ability to evaluate the claims of the authorities of his time and to reject them in favor of his own ideas, not capriciously, but because the ideas of his predecessors were not coherent in the context of what Anaximander observed.[16] In the time of Anaximander, the world was conceived to be divided into solid land (earth), and the sky and heavens. Anaximander's mentor, Thales, had stated that the earth was solidly supported by infinite pillars, and thus below our feet was an endless source of solidity – as it still feels to us when we stand on firm ground. However, Anaximander observed that the Sun set in the west and rose in the east. He also observed that the stars rotated in a pattern so that they dropped below the horizon and then reemerged. If the earth was resting on pillars of solidity that extended below us indefinitely, how did the Sun make it past these pillars in its travel from the western horizon to the eastern horizon, to rise again each morning? Anaximander disagreed with the prevailing theories of his time – he believed the earth must be floating in space so that the Sun could go around it after setting and thus return to the eastern horizon each morning.[17,18]

Anaximander also took the heretical step of removing divine causes as the source for occurrences in the natural world. To Anaximander, natural causes were responsible for natural effects. In making

---

[16] Rovelli C. 2007. *Anaximander*. Yardley, PA: Westholme Publishing.

[17] While Anaximander made great progress in this notion, he based it on the mistaken idea that the Sun went around the Earth.

[18] Anaximander's was not the only hypothesis that was consistent; indeed, even he could have come up with a more complex system that maintained the idea of pillars and had the Sun go from horizon to horizon.

this leap, Anaximander allowed HD coherence to be born, for as pointed out in Chapter 4, if the happenings of the world are at the whim of a capricious and unpredictable deity, then one cannot apply deductive reasoning in understanding how things work. Equally as important is that Anaximander removed the authority of the priestly class and of self-proclaimed terrestrial gods when he ascribed natural causes for natural phenomena. Thus, it could be argued that the intellectual chisel of Anaximander created a fissure that separated natural philosophy (science) from religion.

Of course, that separation is not stable, has been reversed and challenged over time, and remains a constant struggle. Even today there are societies ruled by strong monarchs (or oligarchies) who claim to know the mind of god, who apply their version of god's word indiscriminately, and who punish those who would disagree with that word with torture and death. There are equally authoritative secular regimes. But if there can be no challenge to authority, then there can be no science. It is not surprising that science exists in its most productive forms in free and open societies and is often inhibited or even extinguished in fierce autocracies. This is not to say that fierce autocracies with profound power structures don't support science. The Soviet Union under the rule of Joseph Stalin was a harsh and homicidal regime. Yet Soviet scientists made a large number of contributions, were awarded Nobel Prizes, and advanced technologies in a number of fields. The Soviet Union had rejected the Czar's monarchy as well as divine gods and their religions; they focused (at least in theory) on an atheistic regime of the people.

At the same time, the Soviet Union could not tolerate scientific ideas (or even debate) that had the potential to undermine fundamental assumptions of the communist program. Indeed, Stalin himself, with no scientific background or training, weighed in heavily in scientific debate and dictated the "proper" conclusions of many scientific arguments. Russia faced an agricultural crisis in the 1920s, and in 1928 the propaganda machine latched onto a young Russian agriculturalist named Trofim Lysenko.

Lysenko came from a modest background, was a dynamic young speaker, and tapped into the Soviet narrative of a humble people overcoming the entrenched establishment. Whereas the rest of the world was exploring and developing Darwinian theories of random mutation and natural selection, Lysenko embraced a theory that was an offshoot of Lamarckian theory.

In defense of Lamarckism, it is a perfectly plausible theory arrived at by a retroduction, which was entirely consistent with what was known about biology at the time. Jean-Baptiste Lamarck, a French naturalist, postulated that the behavior of an organism could change the traits inherited by its offspring. This was in stark contrast to Darwinian theories, where populations vary through random genetic variation, and those variations that are most successful at reproducing (in a given environment) become the predominant forms. In Darwin's theories you are born with certain variations, which will give you advantages or disadvantages. In Lamarck's view, you can change your genetics during your life by how you behave and what you encounter.

Lysenko rejected the idea that genes existed at all and claimed that he was able to dramatically increase crop yields, and to even transform one kind of plant into an entirely different plant, based on the conditions in which he grew them. Lysenko was portrayed as a hero to the Soviet people. In 1935 Stalin allowed Lysenko broad and sweeping powers to discredit any scientist (or anyone else) who contradicted his theories. No research or even ideas were tolerated if they went contrary to Lysenko's notions. He was able to discredit his scientific opponents and remove them from their academic positions. Thousands of his opponents were incarcerated and, in some cases, executed.

Lysenko was a wolf in sheep's clothing from the standpoint of scientific dialogue, or any kind of science whatsoever. While he wrapped himself in the trappings of scientific legitimacy, his scientific regime was a precise demonstration of exactly why authority, and the inability to question it, is anathema to any kind of meaningful scientific research or progress. By 1948, Lysenkoism was officially a state-sponsored Soviet theory, and the teaching of any other form of biology was vilified as a subversive, anti-Marxist plot. These prohibitions ultimately spread throughout many of the Eastern bloc countries; in the meantime, genetics progressed rapidly in the rest of the world and the Soviet Union was left behind. After Stalin's death Lysenko slowly lost his influence, although Stalin's successor maintained the views of Lamarckism. Lysenko became more and more marginalized as opponents became emboldened and were able to speak without being victimized. However, it wasn't until the 1960s that Soviet geneticists were free to pursue scientific theories other than Lysenkoism.

While authoritarian societies can encourage technological development, the dictation of what ideas are "correct" and which are "mistaken" by a central authority is inconsistent with modern science. While such an approach does lead to a body of "knowledge," it is a perversion of what science is within free societies. Although Lysenkoism is a classic example of misguided science in an authoritarian regime, it is worth noting that governmental authority has also disavowed aspects of science in societies typically considered more free. For example, there have been repeated legislative attempts in the United States to pass statutes making the teaching of evolutionary biology a crime. Teaching evolution was illegal in Arkansas until 1968 (*Epperson v. Arkansas*). Likewise, the teaching of archeology and geology has been staunchly opposed by religious conservatives because they violate scripture by presenting theories that Earth is older than 6,000 years, as cited in the Bible. This is no less the subjugation of free debate to an absolute authority than was Lysenkoism. The Bible is basically a "because I said so" justification. These debates continue on, in a sad (but predictable) fashion. While there are still those who argue that evolution should not be taught in public schools, more recent arguments center on the view that intelligent design should be taught alongside evolution as a competing scientific theory. To be clear, there is nothing wrong with arguing that the Bible is correct. This is just good scientific process, let alternate theories be investigated and evaluated with how they tie into our web of belief linked to observation of the natural world. The problem isn't in arguing for or against anything, the problem is attempting to silence other points of view (e.g., making it a crime to state it or against the law to teach it) and insisting upon a single view based on authority and not a web of belief at least linked to, if not driven by, observation of the natural world.

So what about the government preventing the teaching of creationism or intelligent design in public school curricula, isn't this the same thing as prohibiting the teaching of evolution? Isn't this exactly what is being argued against? No one has advocated the prohibition of teaching intelligent design as part of a comparative religion curriculum. The problem is teaching intelligent design as part of the science curriculum. Intelligent design cannot be taught as part of a science curriculum, because it is not science. It isn't that intelligent design might be bad science; it is because intelligent design isn't science at all. It has no deductive character between causes and effects and represents a nonrule-governed system (as outlined in Chapters 4 and 5) and as such is not susceptible to HD coherence or scientific methodology.

I am not stating that intelligent design isn't a good theory or that it doesn't explain a lot. Indeed, intelligent design theory explains the natural world much better than does evolution, or any scientific theory for that matter, because intelligent design explains everything that ever has or ever will be observed. Why is the world the way it is? Because an intelligence designed it that way. Perfect, all questions answered, game, set, match – time to take our questions and go home with the warm happiness of complete knowledge. Is it any wonder why this answer is so satisfying to those seeking explanations? But these are not scientifically useful answers; again, this is not science at all because there is no predictive power, no ability to evaluate, and, unlike evolution, there is no evidence that can theoretically refute the theory because there are no deducible outcomes other than the world will exist as the designer wants it to (a detailed analysis of this issue is presented in Chapter 13).

In a misguided attempt to equalize evolution and intelligent design with regards to their scientific status and thus allow intelligent design to be taught in public school science curricula, the school district in Cobb County, Georgia, added the following sticker to public school biology books in 2002:

> "This textbook contains material on evolution. Evolution is a theory, not a fact, regarding the origin of living things. This material should be approached with an open mind, studied carefully, and critically considered."[19]

Cobb County's explanation was that they were trying "to foster critical thinking among students, to allow academic freedom consistent with legal requirements, to promote tolerance and acceptance of diversity of opinion, and to ensure a posture of neutrality toward religion."

At first glance this seems to be a pretty reasonable statement. It seems like a good thing to promote critical thinking – a strong part of science. However, upon reflection, it seems a bit odd to say this about evolution and not about every single other theory being taught in science. Isn't every theory in a science textbook only a theory? Isn't it the case that nothing in a science textbook is an unassailable fact? In large part, the nonfactual status of all scientific "truths" we hold today (until we reject them tomorrow) is one of the major characteristics of science itself, although it has also been argued herein that many lay people (and even some scientists) don't appreciate this. This is the damage that results from not understanding that while scientific knowledge claims may have a huge amount of evidence to support them, they are not "factual" in the way that term is normally used outside of science. In a highly articulate objection to the sticker,

---

[19] August 22, 2002. "ACLU Sues Over Evolution Disclaimers in Textbooks." *Fox News*. New York: Fox Entertainment Group. Archived from the original on March 5, 2016. https://en.wikipedia.org/wiki/Creation_and_evolution_in_public_education_in_the_United_States#cite_note-26.

six parents sued Cobb County, *not* on the grounds that evolution isn't "just a theory," but along the lines stated here. Jeffery Selman, who brought the lawsuit, wrote, "It singles out evolution from all the scientific theories out there. Why single out evolution? It has to be coming from a religious basis, and that violates the separation of church and state."[20]

Nonscientific authority's effects on science are not limited to those of a theological nature. For example, the corporate interests who would be hurt by a decrease in fossil fuel consumption have brought great political pressure to bear against research into climate change in general and global warming in particular. In previous years the cigarette industry assaulted research linking smoking to cancer and emphysema, heart disease, and other maladies. This is not to say that all opposition to global warming or the ill effects of tobacco products are nonscientific. Indeed, social authority of scientific societies is probably just as guilty in recent years of allowing groupthink dynamics to vilify some reasonable objections to mainstream theories. However, this does not negate the fact that authoritative pressure has interfered badly with science.

In 2003, Andrew C. Revkin with Katharine Q. Seelye published an article in *The New York Times* reporting that the George W. Bush administration had edited the scientific content of an EPA report on global warming.[21] Subsequent stories in 2005 confirmed not only the omission of relevant information, but the direct editing of text so as to alter its meaning and impact.[22] In 2006 Mr. Revkin reported that, under government direction, NASA was preventing a prominent climate scientist from sharing findings and conclusions with the

[20] August 22, 2002. "ACLU Sues ..."
[21] Revkin AC, Seelye KQ. June 19, 2003. "Report by E.P.A. Leaves Out Data on Climate Change." *New York Times.* www.nytimes.com/2003/06/19/us/report-by-epa-leaves-out-data-on-climate-change.html
[22] Revkin AC. June 8, 2005. "Bush Aide Softened Greenhouse Gas Links to Global Warming." *New York Times.* www.nytimes.com/2005/06/08/politics/bush-aide-softened-greenhouse-gas-links-to-global-warming.html

public because they differed with the position of the White House.[23] In response to these stories and others, a formal congressional investigation into the matter was launched in 2006, which ultimately concluded: "The evidence before the committee leads to one inescapable conclusion: the Bush administration has engaged in a systematic effort to manipulate climate change science and mislead policymakers and the public about the dangers of global warming... White House officials and political appointees in the agencies censored congressional testimony on the causes and impacts of global warming, controlled media access to government climate scientists, and edited federal scientific reports to inject unwarranted uncertainty into discussions of climate change."[24]

Unlike in the Soviet Union, no scientists were jailed or killed as a result of stating their opinions. However, it appears as though a powerful central authority was modifying scientific conclusions based on a political agenda rather than any notion of scientific work. Is *The New York Times* often considered a liberal paper? Yes, it is. Was the Congress that carried out the investigation in 2006 a Democratic congress? Yes it was. Is it possible that the allegations of interfering were themselves politically motivated? To some extent this seems likely. However, there need be no controversy on the accuracy of these reports and findings, because those individuals in the White House who engaged in the editing process admitted that they were doing so, justifying the activity as "necessary for consistency in meshing programs with policy."[25] While acknowledging that there are politics and bias in all human activities, it nevertheless seems to be the case that the Bush White House was editing the content of expert scientific opinion not based on scientific HD coherence, but on

---

[23] Revkin AC. Jan. 29, 2006. "Climate Expert Says NASA Tried to Silence Him." *New York Times*. www.nytimes.com/2006/01/29/science/earth/climate-expert-says-nasa-tried-to-silence-him.html

[24] U.S. House of Representatives Committee on Oversight and Government Reform. Dec. 2007. "Political Interference with Climate Change Science under the Bush Administration." www.hsdl.org/?abstract&did=481710

[25] Revkin, June 8, 2005.

an agenda born of a political policy that had been formulated according to issues and interests unrelated to scientific work.[26]

The ability of nonscientific authority to interfere with legitimate science depends upon a misunderstanding of what science is, which underscores one reason why the arguments in this book are so relevant to our society. No one is claiming that the Bush White House fabricated data or made up observations that didn't occur. However, what they did do was to alter the strength of the conclusions by removing text (at times whole paragraphs) and inserting select words that had the effect of interjecting uncertainty into the report's conclusion. Phrases that originally referred to "uncertainties" were modified to read "significant and fundamental uncertainties." As another example, "Many scientific observations indicate that the Earth is undergoing a period of relatively rapid change" was altered to read "Many scientific observations point to the conclusion that the Earth *may* be undergoing a period of relatively rapid change." [emphasis added] Similarly, "The attribution of the causes of biological and ecological changes to climate change or variability is difficult" was changed to "The attribution of the causes of biological and ecological changes to climate change or variability is *extremely* difficult." [emphasis added]

Why is it such a big deal to insert these words? Isn't one of the main points of this book that there are no certainties, no absolute knowledge, and nothing about which we cannot be wrong? The problem is that the lay public and even some professionals in the sciences believe that there can be "proof" and certainty, and that anything that contains uncertainty is therefore inconclusive, equivocal, and can be safely ignored until it is proven absolutely true – a logically impossible task. In science-speak, saying that "the Earth is undergoing a period of relatively rapid change" is about

---

[26] It is not my intention to single out the Bush administration. Many, if not all, administrations probably engage in activities of this sort to some effect. Currently, there is much ongoing debate regarding the Trump administration, alternative facts, fake news, and the seemingly tenuous link of evidence to "fact."

as conclusive as one can get. Inserting the word "may" changes a firm conclusion to a speculation. Saying it is "extremely difficult" to attribute climate change to biological or ecological change creates the impression that science is not only unable to make any certain conclusions, but that science is unable to make any conclusions at all, even if imperfect.

There are two main problems here. The first is that a nonscientific body, one that wasn't even attempting to follow scientific norms, was editing scientific findings to twist them to a preexisting agenda. The second problem is that they were able do to so in a way that opponents of the scientific findings could say, "See, the human role in climate change hasn't been proven yet, so we don't need to act." This was able to happen not only because specific statements of uncertainty were added to the report, but also because people have the misbegotten idea that absolute proof is achievable, that such is our standard of proof, and that unless it's achieved we don't have to worry about problems because they aren't yet real.[27]

THE NEED FOR FRINGE BELIEFS THAT PUSH
BOUNDARIES

An important and often underappreciated factor that emerges from the societal dynamic of science is the need for a small (but consistent) population of scientists with unorthodox and fringe beliefs. If everyone who was steeped in the dominant school of thought simply worked within the framework of the existing paradigm, there would be no new theories, and paradigms couldn't evolve over generations. New ideas don't have to be on the fringe, they can be fairly conservative, but conservative new ideas can't break free from existing thought paradigms. Ideas on the fringe challenge the existing web of belief and are essential instruments of ongoing scrutiny of HD

---

[27] For the interested reader, the movie "Merchants of Doubt," based on the book referenced previously, presents this general argument in exceptional terms, using different examples, that illustrate the point with expert clarity.

coherence and the iterative process itself. Of course, one doesn't want the majority of thinkers to be running off on fringe theories, as this may result in academic chaos. Rather, a small percentage of fringe believers is necessary to maintain a sufficiently diverse portfolio of ideas that will allow the advancement of theories over time.

Those individuals who forward the fringe theories may not get any credit for several reasons. First, while they advocate the theory or even invent the theory, they may not be the ones who establish its validity. From the 1500s until Semmelweis's work in the mid-nineteenth century, a series of individuals, including Fracastoro, Kircher, Leeuwenhoek, Andry, Bradley, and likely others, proposed that small entities, too small to be seen by the human eye, caused disease by direct contact. Semmelweis's studies on childbed fever were just another step in a long line of speculation. His interventional experiment (hand washing and disinfection in hospital wards) was ignored and discredited, even though it led to lower mortality rates. His ideas were rejected by the scientific and medical establishment, and Semmelweis died of severe beatings after being committed to an asylum. Later, Snow and Pasteur added more and more evidence to the idea of germ theory until Koch finally "proved it"[28] (even though Bassi had illustrated the principle in silkworms over a century earlier).[29] Science very much needs its fringe thinkers, although it often treats them badly.

AUTHORITY IN SCIENCE

If authority is anathema to science, then one is forced to ask: What is the role of a "scientific authority?" It is ridiculous to believe that there is no authority in science. As much as natural phenomena and data are arbiters of understanding in science, likely to an extent that exceeds most if not all other fields, there are clearly individual

---

[28] This is a highly simplified description of the history, and as such, not entirely correct with regard to details and nuances.

[29] Of course, there were many more players involved in this process, but I as focusing on several of the major players for the sake of simplicity.

scientists who claim authority all the time. Not only do scientists claim authority for themselves, but it is assigned to them by others. Scientific meetings invite "experts" to deliver lectures in their fields. Journals invite experts to write articles expressing their views. There are even journals entitled *Current Opinions in* [insert field of study], based on the idea that the opinions of recognized experts are worth publishing (which they well may be). Since science is a human activity – made up of humans and the social structure from which it arises – there is no avoiding both personal and societal authority at some level. This goes back to the earliest beginnings of science. Anaximander may have been an inspiring figure precisely because he was willing to challenge the authority of Thales, his mentor, who was the greatest authority of the time. But by doing so, Anaximander became the new authority himself.

Authority is by no means reserved to individual scientists. Esteemed groups of scientists (the National Academy of Sciences, Royal Societies of Science, etc.) claim scientific authority as their fundamental function. When professional societies of scientists issue consensus opinions, the weight of their opinions may be based on data, but it is also certainly influenced by the trappings of their expert status as being part of the establishment. The government, the lay public, and scientists themselves use the position of these experts to justify their own positions. The journalist A. J. Liebling once said: "Freedom of the press is guaranteed only to those who own one." While that may have been true, the invention of public media and the Internet has given a printing press to almost anyone who seeks one. However, in scientific publication, peer-reviewed journals are seen as the only legitimate means of publishing bona fide scientific findings. The journals are typically run by an editorial board made up of scientists who are recognized experts in the field and the papers themselves are reviewed by experts in the field, and so expertise-based authority regulates the very nature of scientific currencies.

The prevalence of authority in the sciences might be seen, by some, as contradicting the notion that there is something different between science and any other system. If, at the end of the day, what we take to be true is nothing more than what is accepted by the experts, then what difference does it make if the vehicle of that authority is a religious leader, a political leader, or a leader in science? Isn't it all about authority at the end of the day – just what someone says is true? This is certainly a reasonable objection and concern, and it is imbued with a certain amount of legitimacy. The answer to this question is that authority is a cultural issue of societal dynamics. There are established norms of acceptable and unacceptable societal behavior. These norms of behavior are decided upon by the societies themselves, based on the opinions of current members. These opinions are influenced heavily by historical norms and precedent, but they also change over time. At the current time, in Western science, many agreed-upon rules of conduct and practice are solidly in place, which have been discussed throughout this book. Authority in science has a different meaning than authority in other settings. Scientific authorities are experts in the factual and methodological content of their area of science, and while they must form opinions that are authority based, their authority is restricted (at least somewhat) by the working paradigms of their field, which rest on previous scientific exploration of the natural world. Clearly, the accuracy or meaning of data can be disputed, but one cannot simply change things because they don't fit the outcome one prefers (as did the Bush White House in changing the statements of the scientific bodies). The accepted basis for authority is not "because I said so," but rather rests (or at least should rest) upon ideas derived from observation of the natural world. In this way, it differs somewhat from authority in many other fields.

THE POTENTIAL DESTRUCTION OF SCIENCE BY SCIENTIFIC AUTHORITY ITSELF

Because Western science is a social construct, is defined by scientists, and evolves over time, it contains the seeds of its own destruction, sown into the ground in which it grows. An irony of science is that scientific authorities agree that it doesn't make its final claims based on human authority, but rather on properties of the natural world. However, this is human authority stating in unequivocal terms that one should not accept knowledge claims based on human authority; therefore, such authorities could change their minds and practices. This is what happened in Stalin's Soviet Union in the context of Lysenko's program in heritability. Scientific methods have developed by overcoming errors in normal human thinking, observation, cognition, reasoning, etc. Because our appreciation of human error is continuing to develop, so must scientific method continue to develop. Thus, scientific societies must have the ability to change norms. However, with such power (which is authority based) also comes the potential to destroy itself. Our scientific culture could be lost if the greatest traditions of science are abandoned by subsequent generations (as occurred with Lysenko in the Soviet Union, although it was recovered by groups of scientists when Stalin died and Lysenko's power waned). There is a danger in having a system that is defined by what humans say it is, but there seems to be no way around this. Nevertheless, the danger must be guarded against to the greatest extent possible. This is an ongoing and real-time concern about which we should all be vigilant. The data are what the data are. Societies can choose to act or not act on what scientists observe, but changing the observations based on what we wish were so, or what our political or religious authorities decide should be the case, has always led to far worse situations than accepting and dealing with the data as they are observed.

The danger of scientific authority also goes the opposite way. What if scientific societies become too stringent in their rules for what is "real science"? What if legitimate sources of inquiry become excluded, marginalized, and thus destroyed? For all we know, such is already the case. In his famous 1975 book, *Against Method: Outline of an Anarchist Theory of Knowledge*, Paul Feyerabend argued that if we were to apply the current "rules" of "proper scientific conduct" to the greatest scientific luminaries of history (e.g., Galileo), they would not qualify as scientists.[30] Feyerabend sees the act of defining rules of proper science as a source of group tyranny that serves to crush creativity and new knowledge. He argues that this will destroy science, in a Lysenko-like manner, something akin to the statement: "If you're not doing science the way we say you should, then you will not do science at all!" As John Stuart Mill pointed out in his essay "On Liberty," democracy is not the absence of tyranny; one has simply traded the tyranny of a monarch for the tyranny of the majority. Feyerabend argued that organized and professional science is a tyranny of self-proclaimed "real" scientists that would destroy further progress. In Feyerabend's view, the rules of science should be "anything goes."

From one point of view, Feyerabend may have been a bit misguided for several reasons. First, I don't think it's clear that those who have made the most progress (e.g., Galileo) were the best scientists (as discussed in Chapter 10). They may have been one of many creative scientific thinkers who happened to get it right, and history makes no mention of the myriad creative scientific thinkers who got it wrong. However, if the real stuff of science is done by societies of follow-up scientists in a dynamic between individual and group contributions, then it is quite okay to define science in a way that excludes single luminaries such as Galileo, as they would be the source of the ideas (and some good scientific investigation) but do not themselves define the scientific process. Second, science changes over time, so applying today's standards to the behavior of scientists of previous eras and saying the standard doesn't hold because it would exclude those scientists is not a fair comparison. Rather, scientists should be held to the standards that were accepted at the time when they were working. For example, use of statistics to quantify uncertainty is an essential part of modern science today. However, saying that statistics cannot be

---

[30] Feyerabend P. 1975. *Against Method: Outline of an Anarchist Theory of Knowledge*. London: New Left Books.

an important part of science, because this would exclude Galileo or Newton from being scientists is an unfair criterion. Statistics had not been invented when Galileo or Newton were doing their work. However, Feyerabend's point is important, and we do run a risk of excluding some important knowledge if we reject it out of hand as "nonscientific" due to rules that are not always helpful and at times may be harmful.

At the end of the day, scientific society, like all societies, has its mainstream dogma and establishment views. As one drifts more to the extremes, the views become more iconoclastic and revolutionary, and less those of the establishment. When one gets to the fringes of science, one encounters those who would directly contradict the orthodoxy. Finally, one gets "beyond the fringe" (or perhaps "beyond the pale") to those whose claims can be dismissed out of hand as pseudoscience at best and outright lunacy at worst. Sometimes the fringe thinkers emerge from within scientific orthodoxy and move to the fringe, like Blondlot's N-rays, Peter Duesberg's belief that HIV does not cause AIDS, and Pons' and Fleischmann's cold fusion. These famous figures turned out to be fringe thinkers who were ultimately wrong (at least for now), and in some cases progressed from being fringe to falling right off the playing board. In other cases fringe thinkers come from outside the scientific orthodoxy, as in the case of Semmelweis, who never made it into the orthodoxy and died in obscure ignominy, although he ultimately turned out to be correct.

The orthodoxy of science turns out to be the slow and progressive arbiter of cautious change, whereas the fringe serves as the source of innovative ideas and creative notions. It is not likely that a single individual can be simultaneously imbued with conservative dogmatism and innovative creativity. In fact, some would say (e.g., Kuhn) that this is impossible because observations are theory-laden and, as such, those indoctrinated to a paradigm literally cannot see outside it. To have both, one must have a scientific community that self-regulates. The fringe thinkers are an absolute requirement to prevent the dogmatic orthodoxy from becoming a theocracy. The dogmatic center is required to prevent the fringe thinkers from randomly over-turning established ideas and theories without a firm evidentiary basis for doing so. This is messy, full of controversy, and at times devolves into seeming anarchy, with multiple competing theories all claiming to have the best explanation for existing data. In general, none of the theories explains all of the data (either because the theories are not 100% correct or because the data is not 100% correct, or both, as is typically the case).

While natural phenomena are the final arbiter of knowledge in science, and in this sense rule the day, the generation and interpret-ation of observation is a human activity filtered through human societies, and subjected to an authority structure. So long as science maintains its current character and greatest traditions, so will author-ity be the lesser determinant than is nature. But authority will always be present and will prevail in some cases (especially in the short-term) over the forces of nature's judiciary.[31]

---

[31] If there is any area of scientific scholarship that has a deep and rich content and to which the current work can't even scratch the surface, it is the societal and social component of science. The interested reader is encouraged to avail him or herself of the well-developed literature in this area. At the very least, one must understand that science is not a cold body of logic around clear observations of nature, it is a hot swirling social construct in which the very observations of nature themselves and all the thinking about them are affected by societal factors. Some have even argued that science is first a social construct, as all other component parts are generated with strong societal influence.

12 A Holistic World of Scientific Entities, or Considering the Forest and the Trees Together

In 1924, a South African named Josephine Salmons made a visit to the home of Pat Izod, a family friend. She noticed an odd, humanlike skull sitting on his mantelpiece. Curious, she asked him its origin and learned that it had been found by a miner working at the Buxton Limeworks. This miner wasn't focusing on questions of human origin nor was he testing any particular hypothesis; rather, he was blasting through limestone in an effort to increase the output of the mine. He was no different than someone who goes out for a walk and notices an interesting tree or is taken by the shape and glimmer of a particular puddle. He noticed the skull and gave it to his employer, E. G. Izod, who was a visiting director of the Northern Lime Company, which managed the mine. E. G. Izod gave it to his son, who put it on his mantle. Josephine Salmons happened to be a young graduate student working in the laboratory of Dr. Raymond Dart at the University of Witwatersrand in Johannesburg. Dr. Dart was an anthropologist of Australian origin who had taken the position of professor two years earlier.[1]

Recognizing the potential importance of the skull, Salmons brought it to Dr. Dart, who shared her enthusiasm and interest. Dart then contacted the mining company and launched an investigation into additional fossils and skulls present at the mining site, leading to a body of work that defined a previously unknown ancient hominid species that they named *Australopithecus africanus*, which lived in the range of 3 million years ago. Called by some "the missing link," this discovery supported Darwinian theories of evolution. For those who doubted "creation stories," this was further evidence against all

---

[1] Jan. 25, 2019. "Raymond Dart." *Wikipedia*. https://en.wikipedia.org/wiki/Raymond_Dart

existing life having been created exactly as it is today. The accidental observations of a limestone miner wound up as essential data relevant to fundamental scientific hypotheses.

In 1964, two scientists working at AT&T Bell Labs in New Jersey, Amo Penzias and Robert Wilson, were trying to detect radio waves that they were attempting to bounce off of balloon satellites, a practical exercise in engineering new technology. As part of a typical problem-solving process, they became aware that to detect very faint signals they needed to eliminate all background noise and interference from other sources. After taking painstaking measures to eliminate interference from local broadcasters and the uninvited pigeons that were nesting in their detection dish, Penzias and Wilson concluded that there was a faint but persistent signal of background noise coming from sources other than Earth; in fact, the signal was coming from outside our galaxy and appeared to be ubiquitous. This was undoubtedly disappointing from an engineering perspective, as there was no way to eliminate this "background radiation." However, their finding served as fundamental data in the scientific evaluation of two competing hypotheses about how the universe came to exist.

The Big Bang theory[2] had posited that the universe exploded forth from a single compressed point. In contrast, the Steady-State theory posited that the universe has always been in existence. The Big Bang theory predicts that there should be residual signals throughout the universe – basically, an echo remaining from the Big Bang. In contrast, the Steady-State theory leads to no such prediction. Thus, the observation of background radiation throughout the universe supported the Big Bang theory.

A miner at the Buxton Limeworks saw a skull he thought was cool. Amo Penzias and Robert Wilson were frustrated at being unable to eliminate background radiation so they could detect transmitted signals. None of these individuals was performing an experiment to

---

[2] Here I am not referring to the popular TV show, although the characters on that show clearly believe in the theory of the origin of the universe after which their show is named.

test hypotheses per se, and yet their work resulted in the testing of the predictions of fundamental scientific hypotheses, upon which they were not working and of which they may not have been aware. This type of occurrence happens over and over again, precisely because of the holistic nature of knowledge – the interlinking of our wide web of belief in an interconnected universe. This is no different than if you were digging a hole in your backyard to plant a new blueberry bush and happened upon a dead body that turned out to be Jimmy Hoffa's remains.[3] By the simple act of gardening you would have generated data to solve one of the mysteries of American history, but you would have done so without any intention to solve this mystery, and potentially without even knowing who Jimmy Hoffa was or that there is a question of what happened to him.

Scientists frequently carry out studies for a purpose and fail to achieve their goals, but instead succeed in testing the prediction of an unrelated hypothesis. Sometimes this is obvious to the scientist, as it is in their field of study, but in many cases, they have generated data that tests a prediction of which they have never heard, and of which they are unaware until they publish their findings, and someone else realizes its importance to a different area of study. This doesn't sound much like the linear and logical description of scientific method that many hold to be true. Rather, it sounds a bit random, because it is.

PRACTICAL EFFECTS OF A HOLISTIC WORLD OF INTERRELATED THINGS (THE ONENESS OF NATURE)

Ideas of the "oneness" of nature or the holistic sense of the universe – of the interconnectedness of all things – is more often associated with spiritual systems of belief and philosophy than with descriptions of science. Yet our best understanding of science indicates that the world is interconnected, that one cannot make a hypothesis about one thing without making a hypothesis about many things, and likely

[3] Anyone young enough not to know who Jimmy Hoffa was can google his name.

all things, and that at the end of the day, the world consists of a series of related entities and phenomena.

On the surface, science seems to go against the very grain of a holistic universe. If nature is really one thing, then scientists spend an awful lot of misguided time dividing this one holistic world into smaller and smaller parts, and defining these parts as more and more separate things, to the point that one might think the entire natural world was sorted into silos between which there is no communication at all.

Biologists divide life into taxonomic domains – into kingdoms of animals and plants, into phyla, classes, orders, families, genera, and species. Chemists divide all matter into a table of elements made up of atoms, which combine according to certain rules. Physicists further divide atoms into electrons, neutrons, and protons. Protons and neutrons can be divided into quarks. Electrons are thought to be elementary, but they have emergent properties when interacting with other matter, in which they manifest as holons, spinons, and orbitons. Anatomists divide humans into organ parts, histologists divide these organs based on cells and cell type, and biochemists divide cells into lipids, proteins, nucleic acids, and carbohydrates. Even psychologists divide the human mind into component parts (e.g., ego, id, and superego). So what about this whole system seems to be holistic and of one thing?

Division of the world into smaller and smaller units is a necessary thing, at least in the context of scientific methodologies, in order to make any progress in understanding the natural world. This may not be necessary for other systems of belief to process experience; in fact, doing so may be anathema to other systems of belief and render them meaningless. However, science is largely interested in the association of things, in the causes of effects, and of how one type of thing influences another type of thing. In order to think in this way, one must have categories of things to study.

Science tends to focus on small, isolated areas of the natural world, one at a time. Eventually, science attempts to integrate the individual pieces back into a broader picture. However, focusing on small, isolated areas is a necessary part of scientific investigation. The reason for this can be found in the framework of holistic coherence, as described in Chapter 3. Any hypothesis can be rescued from seemingly rejecting data (things it didn't predict) by evoking auxiliary hypotheses and unobserved entities. Thus, to make any progress, one must control as many auxiliary hypotheses as possible, in an attempt to single out the one thing you wish to test. This is never completely possible, but one can certainly do better than passively observing an uncontrolled natural world. Imagine you were attempting to solve a one thousand-piece jigsaw puzzle, but instead of working first on the border or some part of the picture you had to work on the entire puzzle simultaneously. Because you couldn't focus on the interaction of two single pieces, you'd probably never make any progress. By focusing on individual sections you would still make errors, as certain pieces will at first look like they're going to fit together but then don't. Other pieces won't appear to fit together until you change their orientation, but ultimately, they will fit together. By working on isolated parts of the puzzle you can eventually fit the small parts together to create a greater, integrated whole.[4]

The societal part of science, discussed in Chapter 11, has the dynamics of a jigsaw puzzle in which multiple groups of people are simultaneously working on different parts of the puzzle, stealing pieces from each other, but then discarding them when they don't fit. Moreover, several groups of people are likely to be working on the same part of the puzzle simultaneously, making errors and progress, undoing errors, but also undoing progress others have made as well by swiping pieces and at times sabotaging each other. None of this means that there isn't an overall picture contained in the whole puzzle; however, none of the assemblers knows what that picture is because

---

[4] Many others have used a variety of analogies to puzzle solving to describe science, most notably Thomas Kuhn in his seminal work, *Structure of a Scientific Revolution.*

there is no box cover to look at. Nevertheless, the overall picture is there; it just hasn't yet been perceived.[5]

Despite the requirement of categories to construct webs of belief, the intrinsic interrelatedness of things can be found in many of the characteristics of science described thus far. Indeed, a lack of appreciation of this interrelatedness has been responsible for much confusion and a lack of understanding of science by scientists themselves. One could look at the raven paradox (Chapter 3) as simply a statement that a hypothesis about one thing is actually a hypothesis about all things. One cannot make a statement about black ravens without also commenting on white ravens, yellow school buses, and colorless fish. Of course, most individuals or groups will never have the perspective that what they study is related to all other things or even to most other things. Who among us can see more than just a miniscule portion of the puzzle that is the universe?

In 1828, Friedrich Wöhler was a 27-year-old teacher in a technical school in Berlin. While this position afforded little in the way of income or esteem, it did provide Wöhler with a laboratory in which he could carry out chemistry experiments. One day he was attempting to combine two salts (potassium cyanate and ammonium sulfate) in a way that would produce ammonium cyanate, but the result was something different. Wöhler had inadvertently stumbled onto the conditions that produce urea. The finer details of what preceded and followed this observation are a subject of some controversy, because a number of different histories have been put forth (the most popular of which clearly seems apocryphal); nevertheless, the outcome that Wöhler observed has had a number of important implications. Indeed, who could have imagined that synthesizing the main constituent of human urine would have had any impact at all?

First and foremost for Wöhler, at least according to what he emphasized in his report, was the understanding that both ammonium cyanate and urea had the same proportion of different elements (in this case carbon, nitrogen, oxygen, and hydrogen). Thus, ammonium cyanate and urea

---

[5] The analogy of science to this kind of puzzle solving was popularized by Thomas Kuhn in *Structure of a Scientific Revolution*.

seemed related in some way by their atomic content. Yet, the properties of ammonium cyanate and urea were clearly different. Thus, the properties of the chemical had to be determined by something other than simply the relative abundance of different atoms – it suggested some structure. While this is what had the most meaning for Wöhler, his experiment also had great meaning for an entirely different area of theoretical work, regarding a theory called "vitalism."

Vitalism was a combination of different theories and ideas that were united by the belief that there is a fundamental difference between living things and nonliving things with regard to their composition. At least as far back as the ancient Egyptians, the idea that some life force distinguishes a living thing from a nonliving thing had been posited. However, in a more modern form and in the context of the chemistry being studied by Wöhler, vitalism was centered on the types of chemicals that make up living things and nonliving things. It was well understood that many clearly nonliving things (e.g., crystals and salts) could be heated to high temperatures and melted, but when cooled they returned to their basic state. In contrast, chemicals derived from living things were often destroyed in some way by heat, so that when they cooled they decomposed. This gave rise to the notion that living things had chemical properties distinct from nonliving things. In general, the terms "organic" and "inorganic" were applied to chemicals that were derived from living and nonliving things, respectively.[6]

One of the central ideas of vitalism is that organic chemicals can only be made from living things (often attributed to the need for some life energy). Thus, hypothetico-deductive (HD) coherence of vitalism would predict that one cannot combine inorganic chemicals and have an organic chemical result, in the absence of a living thing to transfer life-energy to the organic chemical. However, in retrospect, this is precisely what Wöhler's synthesis of urea had accomplished. The synthesis

---

[6] The term *organic* has a different modern meaning, which is something akin to "natural and/or nonartificial" to the lay public. To chemists, organic typically means carbon-based or containing carbon, whereas inorganic refers to noncarbon-based compounds.

of urea dealt a difficult blow to vitalism theory, because it goes directly against predicted outcomes. Regrettably, the story itself, and in particular the sequence of events that followed, have been badly modified and/or embellished. As has been pointed out by the historian of science, Peter J. Ramberg, some accounts of the story, "ignoring all pretense of historical accuracy, turned Wöhler into a crusader who made attempt after attempt to synthesize a natural product that would refute vitalism and lift the veil of ignorance, until 'one afternoon the miracle happened'."[7]

There is little evidence that Wöhler set out to attack vitalism in any way, although he was clearly aware of the potential implications of his finding after it had occurred. However, he was as much, if not more, focused on the implications regarding isomers of different chemicals. Moreover, Wöhler's findings were in no way a death knell to vitalism theory at the time he published. To begin with, it was unclear that Wöhler's starting materials were purely inorganic, and the auxiliary hypothesis that his salts had been contaminated with living matter was evoked to rescue vitalism from his findings (an example of holism and underdetermination). However, Wöhler's discovery did provoke ongoing work to generate organic compounds from nonliving sources, and over time, more and more organic compounds were synthesized from inorganic starting materials. Hermann Kolbe made an organic compound starting with the elements in coal. There was no sudden epiphany or "clean moment" where vitalism was rejected. It had been noted by Louis Pasteur that organic chemicals from living things came in only a single orientation (stereoisomers), whereas those made in chemical synthesis had equal proportions. This led to an idea that suggested a vital force was responsible for asymmetry, an idea that lasted well into the 1900s.

Slowly, as the society of science worked this over, and as new generations of chemists came along who were less devoted to the older paradigm, vitalism was rejected as a concept. The current view is that living things are made of the same elements as nonliving things, and it

---

[7] Ramberg PJ. 2000. "The Death of Vitalism and the Birth of Organic Chemistry." *Ambix* **47**(3): 170–95.

is simply the way that they are combined, and not some life force, that gives rise to organic chemicals.[8] This example illustrates how science often does not advance by means of clean and unequivocal experiments based on a deduction from a hypothesis, after which ideas are soundly rejected because nature was not consistent with the predictions. History, however, is often written as though this was the case.

Much more important to this discussion is how experiments designed to test one question (could ammonium cyanate be made) sometimes lead to a result that serves as strong evidence to test a completely separate question (the idea that organic compounds could only come from organic sources). This occurs precisely because of the holistic component to the world. Ultimately, all natural phenomena are related to each other in some way. Wöhler had a hypothesis that predicted the synthesis of ammonium cyanate from two inorganic salts under certain conditions – his prediction was mistaken and urea was made instead. Simultaneously, vitalism theory was a hypothesis that predicted the inability to synthesize of urea from inorganic salts. When two hypotheses each predict an outcome of the same experiment, a person working in one field can inadvertently provide evidence for an entirely different area of study, even if the investigator is unaware of the other field and its theories. Indeed, because of the interrelatedness of natural phenomena, this happens all the time. The person generating the data will likely not appreciate or understand the implications of his or her work to other areas, as it is not on their intellectual radar screen. It is for this reason that the publication of scientific findings and their distribution and availability to a broader audience are such essential acts.

The implications resulting from the holistic nature of the universe and the interrelatedness of ideas have their strongest effect when scientific efforts focus, somewhat myopically, on the immediate problem to be solved. In the funding of biomedical science by

[8] Please note that it is a separate issue whether there is a "soul" with which humans are imbued or if our self-awareness comes solely from the combination of atoms in our brains (a more materialistic view). In the context of this chapter, we are only discussing the idea that the building blocks of animals can be made synthetically and do not have to be derived from another living animal.

both governmental bodies and private foundations, one is often presented with an intense focus on curing illness and helping to mitigate human disease. It seems appropriate, even self-evident, that this should be the case. Grant applications seeking to test the basic workings of biology (without a focus on disease) are often frowned upon as being nonapplied, curiosity-driven science. Ridicule for this type of research has been expressed at congressional hearings, where legislators ask federal agencies to explain why taxpayer dollars are being used to study obscure things. However, because of holism, this can be a damaging view. Focused efforts to cure disease are certainly important; however, one must take note of how often the breakthroughs in one field are the result of information generated in a seemingly obscure area that had an unrecognized but nevertheless overlapping retroductive space.

Arguably the greatest recent breakthrough in medical biology has been the discovery of a new means of modifying the genetic makeup of animals and plants in a highly focused and specific manner. This breakthrough (called CRISPR/CAS9) was discovered not by scientists trying to figure out how to modify genetics to cure disease; rather, it was discovered by basic bacteriologists who were curious about why certain bacteria had long repeated sequences in the ends of their chromosomes.[9] As another example, in 1949 at Oak Ridge National Laboratories, biologists who were breeding mice noticed animals with a weird appearance. They were curious why the mice (called scurfy mice) looked funny, so they began to breed and analyze them.[10] Eventually, their studies identified the responsible gene, and in doing so, they uncovered an entire area of biology that regulates when a person's immune system attacks one's own body (autoimmunity) and when it doesn't, leading to an essential breakthrough for a wide variety of human diseases. One must wonder

[9] Horvath P, Barrangou R. 2010. "CRISPR/Cas, the Immune System of Bacteria and Archaea. *Science* **327**(5962): 167–70.

[10] Ramsdell F, Ziegler SF. 2014. "FOXP3 and Scurfy: How It All Began." *Nature Reviews Immunology* **14**: 343–9.

how well a grant application would be received today by scientists who just wanted to figure out why certain mice looked funny.

Many of our major medical breakthroughs have come from basic biologists grinding up fruit flies and microscopic worms simply because they were curious about odd characteristics they had noted. Why would such seemingly obscure studies lead to important medical breakthroughs in humans? The reason is that all life on Earth is interrelated. It doesn't really matter if it all evolved from common ancestors or if it was designed by a creator. There are common themes and clear relationships between terrestrial forms of life. Thus, the study of the biology of anything often has some relationship to human biology, even if it is not the one intended when the study was conceived or carried out. This is all the more reason why science must be published, as those who make an observation are, more often than not, unaware of the multitude of hypotheses that their observations are simultaneously testing.

This type of dynamic is in no way limited to scientific hypotheses. Consider a group of detectives in Miami who were attempting to solve a murder case. They had DNA evidence indicating who the perpetrator might be, but the DNA profile had no hits in the national database, and the case remained unsolved. Some years later, detectives in Detroit were trying to solve a rape case. They had several suspects and asked each of them to voluntarily submit to a DNA test. Every one of the suspects consented to the test and agreed to give DNA, which was analyzed. None of the DNA matched, effectively ruling out those suspects in the rape being investigated. However, when these DNA profiles were added to the national database, one of the DNA profiles that didn't match the rape case in Detroit came up as a hit for the murder in Miami (now a cold case that no one was actively investigating). It was flagged by the system, and Miami detectives looked into the person who came up as a match. As it turned out, he was living in Miami at the time of the murder and worked with the victim. Further investigation revealed that he had a motive, and a search warrant was obtained for his apartment, where detectives recovered a gun. Bullets fired from this gun matched the ballistics of the bullet taken from the victim. This crime was solved by an entirely unrelated investigation by detectives who were not even aware of the crime they ultimately cracked.[11]

One of the most difficult things for scientists and lay people to accept is how often the greatest scientific breakthroughs come from unlikely sources. It is not just the serendipity of an accidental discovery (as with Alexander Fleming and the discovery of penicillin); rather, it is that a finding in one line of study inadvertently tests the HD predictions of an entirely different line of study. This occurs precisely because, at the end of the day, the natural world is one big thing – an interrelated web. In an effort to break down the world into pieces of a manageable size, we make up categories of things. Stars are different from planets, which are different from comets. Animals are different from plants, and both are different from rocks. Yet the iron that is a required substance for almost all terrestrial life, including humans, and that is a major component of some rocks can only be formed (as far as we know) within the center of a star.

When one tests a particular hypothesis, one is simultaneously testing endless numbers of other hypotheses of which one is unaware. It is for this reason that insisting that science be driven only by goal and purpose is a bit misguided. Goal and purpose are essential to help guide the expenditure of limited research resources. At the same time, goal and purpose are limiting when one considers where advances have historically come from, due to the holistic web of belief. Focusing too much on studies specifically designed to solve a particular problem, with too little support for basic curiosity-driven science, will only serve to significantly slow, or even stop, progress on the very problems we wish to solve.

THE WEB OF BELIEF AND PASSIVE OBSERVATION AS SCIENCE

Quine's "web of belief" indicates that while we may study small, isolated parts of the world, knowledge is actually an incredibly intricate and

[11] I have used a fictional case in order to not have to reference the names of an actual rape or murder victim; however, there have been a number of actual stories like this one.

complex web of interacting ideas and their relationships. This web is by no means two-dimensional, like a spiderweb drawn on paper; rather, it is multidimensional, with complex nodes and a mishmash of interactions between points.

One of the most important implications of knowledge being a series of interconnected ideas that form a web of belief is that the modification of any one part of the web simultaneously alters many other parts of the web. People may not always notice the connections, but they are there. When Friedrich Wöhler modified his small part of the web of belief by synthesizing urea, he inadvertently altered the entire field of vitalism theory. Penzias and Wilson observed background radiation, modifying a very small part of the web around a practical engineering issue and, in doing so, altered part of the web around the very origins of the universe. When Priestley modified his part of the web of belief by observing candles burning and mice living longer in dephlogisticated air – a thing that in retrospect didn't even exist – he changed our knowledge of elements, of combustion, of heat theory, of chemistry, and of life itself.

A SERIES OF DIFFERENT POSSIBLE WEBS OF BELIEF: THE PROBLEM(S) OF CATEGORIZATION AND MISCATEGORIZATION

In antiquity it seemed obvious that there were natural divisions of things in the world, that nature could be carved where "the natural joints are."[12] However, it is currently unclear if that is the case, whether the categories into which we put things are natural or arbitrary. We divide humans into males and females, into races and ethnic origins, into religions and political parties, and we make generalizations based on these divisions. You will find popular literature replete with books about why men act one way and women act another. Regrettably, you will also find much speculation about why people of different races have certain characteristics; or why Christians are

---

[12] Plato, Phaedrus 265e; Fowler HN (Trans.). 1925. *Plato in Twelve Volumes, Vol. 9.* Cambridge, MA: Harvard University Press; London, William Heinemann Ltd.

different from Muslims, Buddhists, Jews, and others; or why people from New York are different from people in Atlanta or Seattle.

Yet even something as obvious as the difference between male and female is unclear. How do you define a male vs. a female? Is the definition based on appearance and anatomy? Humans can be born with ambiguous genitalia, where they have the characteristics of both genders simultaneously. One might be tempted to assign such individuals to the male or female category based on their chromosomes, but while most males have an XY and females have an XX, there are also humans who have XO, XXY, and XYY, among other variations. To what sex would you assign those people? And what about a person's gender identity? To what sex would you assign someone who is male bodied but female identified, and vice versa? What about a gender nonbinary person? This is not to say that there is no difference between males and females, but there is a fuzzy middle, and it is unclear that there is a satisfactory definition of male and female that can be applied to all people.

Centuries of thought and writing have been devoted to discussing the difference between "races." These findings have made their way into laws and liberties (or the lack thereof), based on a person's race. However, it's not clear that there is a biological basis for race; race appears to have more of the characteristics of a social construct. This is not to say that a "black" individual from Africa will not appear different from a "white" individual from Northern Europe or an "Asian" individual from China, nor is this to say that physical appearance isn't affected by the DNA a person carries. Rather, the question is whether there really are meaningful divisions that can define race, if race is a human concept or an actual division in the natural world. Whereas most people will not be surprised that there is debate about race, fewer people will know that there is just as much of a lack of clarity around species. How can this be? Isn't it clear that humans belong to a different species than ferrets? As with sex, in some cases there are clear differences; however, the question is not whether the categories have meaning, but whether the divisions between

categories are clear and definable, and have any fundamental anchoring in nature as opposed to being categories invented by humans.

An even more troubling notion is that there is no clear definition of what it is to be human. If being human is a function of having
the physical form typically attributed to humans, then people missing
limbs are not human. If being human is having 46 chromosomes of a
certain composition, then people with chromosomal abnormalities
are not human. If being human is a function of having a certain
cognitive capacity and self-awareness, then people in comas are not
human and very smart computers may be. If being human is being
able to reproduce with another human to make a new human, then
many people walking around today would be not categorized as
human.

These problems with categorization are not just a linguistic
trick or an example of philosophical analysis run amok. The world
can be categorized by different methods and schemes, and the web of
belief will look very different depending on what scheme is used.
Moreover, moving past the notion that knowledge may at times be
particular and even arbitrary, much damage is done by assigning
things to the wrong category, by having too few categories, and by
having too many categories. Much of scientific progress cannot occur
until categorization schemes are altered.

As described in Chapter 4, the predominant theory of infectious
disease used to be miasma theory, by which "foul vapors" transmitted
disease. When germ theory developed and disease was attributed to
microbes that could be seen under a microscope, there was a logical
and powerful objection to this newfangled notion. Microbes were
essentially everywhere. Bacteria from all humans could be cultured
and seen under the microscope, regardless of their state of health or
disease. If microbes caused disease, then why weren't all humans
sick? This would seem to be a death blow for germ theory – the kind
of rejecting evidence that can be very powerful. The response was to
form a categorization scheme of microbes, i.e., there are different
categories of microbes, those that make humans sick and those that

don't. Moreover, different diseases are caused by different microbes. Cholera is caused by one microbe, leprosy by another microbe, tuberculosis by a third, etc. This categorization goes on today. Different strains of *Escherichia coli* cause profound diarrheal illness or not, depending on whether they happen to express a gene product they picked up from *Shigella*.

We often hear about the search for the cure for cancer, but cancer is actually a large number of different diseases. The categorization of cancer into different subtypes is an entire field of study, and there are multiple categorization schemes, many of which compete with each other and upon which there is no unanimous agreement. The different categorization schemes are compared, and the ones that give the best predictive power of prognosis and/or are best predictors of what therapies are likely to work are favored from a pragmatic point of view. This is one of the major uses of a categorization schemes. However, it is not clear that one categorization scheme is more based in nature than the other.

Although necessary for navigating the world, categorization schemes cause a number of serious problems that afflict both science and normal thinking. One major problem is the lumping together of two things that are actually different. Consider the previous example of *E. coli* bacteria. It is currently understood that there are different strains of *E. coli*. Those strains that have acquired a gene from *Shigella* and produce Shiga toxin cause horrible and at times life-threatening diarrhea. Those that don't carry Shiga toxin generally don't cause diarrhea, and contribute to normal digestive health. If we go back in time before this categorization was appreciated, we would just have a bacterium called *E. coli*. Now, consider an outbreak of diarrhea found only in people who ate sandwiches at a particular buffet. If biologists tested the food at that buffet, they would just find *E. coli*. *E. coli* is essentially in everyone's gastrointestinal tract as part of their normal flora. The biologists could effectively rule out *E. coli* as the cause of the outbreak, because it couldn't be responsible for making people sick if it is found in everyone. They would have to seek another "non-*E. coli*" microbe, which they would never find. The reason they would never find it is because it's hidden by the act of categorizing two different things into one group.

The opposite problem occurs when categories are created that have no measurable outcome on effect. Scientists spend huge amounts of time testing whether certain strains of germs cause disease vs. other strains, spend huge amounts of time testing if one category of cancer responds better to chemotherapy than other categories, etc. However, if they have made separate categories based on some criteria that doesn't lead to a meaningful distinction, or that doesn't really exist due to mismeasurement of the defining characteristic, a large amount of resources is being spent in chasing phantoms.

How one categorizes the world (into what buckets one puts different or seemingly different things) changes the structure of our web of belief and has widespread effects, altering how we interconnect concepts and develop ideas. Many different kinds of webs are possible depending on how we categorize things, and thus the same data from the natural world can have many different meanings as a result. It is unclear whether categories actually exist as a property of nature or if they are entirely human constructs. Therefore, how we should categorize things (or not categorize them) is not abundantly clear. What *is* clear is that we must categorize things to make the world susceptible to analysis. How we do so will have widespread ripple effects on our entire web of belief, and thus how we view the natural world. It is worth considering a pragmatic view here. Since different categorization schemes give rise to different webs of belief, certain categorizations may be highly predictive for some scenarios, whereas a different category-based web would be more useful for a separate scenario. If one focuses on the goal of predicting and controlling nature and sets aside (for a moment) the ambition of finding an absolute and unassailable truth, then changing the category schemes as needed may be a useful thing. However, this may remain fundamentally unsatisfying to most humans (scientists included), because it seems we do not easily relinquish our viewpoint that the categories we view the world to have, even those retroduced for things we have never directly observed, really exist outside of the abstract conceptions of the human mind. We have perceived these categories, we have felt them to be true, and so they must really be.

13 Putting It All Together to Describe "What Science Is and How It Really Works"

SCIENCE ENTAILS THE PRACTICE OF COMPENSATING
FOR ERRORS OF NATURAL HUMAN THINKING
AND IT EVOLVES OVER TIME

Based on the discussions in this book, the following definition of science is suggested to my fellow scientists and nonscientists alike. First and foremost, science is an outgrowth of normal human observation, reasoning, conclusion, and prediction. Scientists and nonscientists both depend upon induction and the assumptions it entails – assumptions that are imperfect and don't always hold. They assume that the future will resemble the past to a greater extent than by guessing alone, and they also assume that what one has encountered today is more representative of things not yet encountered than can be arrived at by random guessing. Both scientists and nonscientists retroduce causes for the effects they observe, a form of reasoning that suffers from the fallacy of affirming the consequent. As a result of this fallacy, scientists and nonscientists both retroduce hypotheses of causal things that likely never existed, such as phlogiston being the cause of heat, a vital force being required for the types of chemicals that come from living things, and the great Sananda causing a prophet's pen to write. One needs ongoing observation, and if possible experimentation, to further assess which retroduced causes one should hold onto (at least for now) and which should be rejected (at least for now). Scientists and nonscientists both use deduction (or at least a form of reasoning that resembles deduction but may not adhere to strict standards of formal logic) to make further predictions based on their retroduced hypotheses. Scientists and nonscientists both have fallacies in their hypothetico-deductive (HD) thinking, make mistaken observations, have cognitive biases, and fall in love with their hypotheses, noticing observations that confirm and ignoring observations that refute. Scientists and nonscientists are both susceptible to social pressures, social biases, and manipulation (intentional and unintentional) by the groups and societies in which they find themselves.

If science and nonscience have so much in common and are so highly related, what can the distinction possibly be? The proposed distinction is that science makes particular note of the source of these errors and develops its methodology (over time) to mitigate these errors. Epistemologists and logicians have continued to expand our understanding of the strength and weakness of inductive and deductive logic and retroductive thinking. They have explored issues of causality and the extent to which it may or may not be testable. They have analyzed problems of affirming the consequent, of underdetermination of theories, of problems of evidence and confirmation. In short, the strengths and weaknesses of our reasoning process has and continues to be analyzed by scholars of human thinking. As problems are brought to light, science modifies its methods and processes to compensate for the newly recognized problems.

As cognitive errors and human biases have become more and more apparent, science has adopted methods of decreasing bias (blinded studies, randomized controlled trials, etc.). As our understanding has evolved of how often apparent effects will occur by chance alone, science has adopted statistical approaches to decrease chance-based errors and to quantify the likelihood of error in any particular setting. The nature of scientific societies, their meetings, and their communications has evolved in a way that helps mitigate the negative effects of excessive individual authority and individual bias. The rules of publication and presentation of data, and the requirement to formally declare conflicts of interest are helping to address hidden biases due to affiliation with different groups and the potential to benefit therefrom. None of these efforts to mitigate error and bias are perfect, but continuing to identify previously unknown sources of error and the aspiration to better mitigate all error is intrinsic to science in a fundamental and defining way. Scientific practice will likely never achieve the goal of eliminating error, but it should always seek to do so with persistent vigilance.

If we define science as a thing that progressively attempts to compensate for errors found in how humans normally navigate the world, it is easy to see why one cannot define what science is by looking for particular common methods found in all scientists (and their work) across the centuries. Much of ancient and medieval science developed an understanding of the fallacies of logic, and they used this new tool to remedy errors in reasoning. However, while they developed a formal system of logic, they didn't fully appreciate that humans don't have much (if any) intrinsic ability to determine base axioms – although humans are quite good at feeling like they can. Moreover, they didn't seem to fully appreciate the underdetermination of retroduction, or if they did, they were not terribly concerned with it. Thus, they developed whole systems of understanding nature without rigorous checking of whether predictions deducible from their belief constructs held. Formal experimentation was little described by the ancients. Scientists in the sixteenth and seventeenth centuries gave greater notice of the need for empirical observation of the natural world and formal experimentation became the norm; however, they likely didn't understand the extent of the difficulties of associating cause and effect. In the seventeenth and eighteenth centuries, controlled experimentation to isolate associations and assess causality became more appreciated, and experimental science began to remedy errors of previous generations, despite having little to no understanding of rates of error and probability theory.

As a case in point, testing statistical rates of error might be seen as one defining characteristic of modern science, or at the very least a common scientific practice, but it was seldom found in scientific work prior to 1900. Does this mean that eighteenth and ninteenth century scientists weren't practicing science or weren't doing it correctly? This would seem a difficult claim to justify. This problem is solved if one defines science as an evolving system that progressively modifies its processes over time, as more and more sources of error with normal human thinking become evident. In this way, science from antiquity to the present day is the same thing – a focus on refining natural human thinking to compensate for errors we make. And the more we become aware of our errors, the more scientific method modifies itself accordingly.

An example of the evolution of science over time can be found through analysis of the paper in which Priestley first isolated oxygen.[1] Clearly the isolation and demonstration of a new element of nature, one that is required for much life on earth, would count as meaningful scientific progress.[2] Yet an examination of Priestley's paper reveals a work that would never be accepted in a scientific publication today. The instruments used and the procedures performed are only vaguely defined, the results are crude and qualitative, and no consideration is given to statistics, bias, etc. He describes having isolated different types of air, and, in particular, one that is "five or six times better than common air, for the purpose of respiration, inflammation, and, I believe, every other use of common atmospherical air." Rather than providing any meaningful quantification by today's standards, he simply states that "a candle burned in this air with an amazing strength of flame; and a bit of red hot wood crackled and burned with a prodigious rapidity..." He went on to write: "I introduced a mouse into it; and in a quantity in which, had it been in common air, it would have died in about a quarter of an hour; it lived, at two different times, a whole hour and was taken out quite vigorous..." Priestley went on to say that he did repeat this finding with

---

[1] Priestley J. 1775. "An Account of Further Discoveries in Air." *Philosophical Transaction* **65**: 384–94.

[2] Priestley was unaware of the magnitude of his discovery and never considered that he had isolated a new element; however, he was clearly excited by the properties of what he had generated.

one additional mouse. It was basically a semianecdotal narrative. This paper would not come close to rising to the minimum standards of science today. Yet, just over 240 years ago, this was about the best chemistry going on in the world. Arguably, it remains one of the most influential discoveries in the history of science.

Scientific methodology continues to evolve. Logicians and mathematicians continue to define the limits of what logic and rational thought can achieve, new scientific instruments and methods are rapidly developing, statistical theory is a progressive and dynamic field, human cognitive psychology continues to uncover biases, and anthropology and sociology continue to refine our view of how human interactions affect human beliefs. I have little doubt that Priestley would view science as it is being done today to be bizarre and unintelligible, not just with regard to content (much has been learned, and forgotten, since his time) but also with regards to process. This process of scientific refinement and reinvention continues today. Unless things go terribly wrong or we lose our way as a society, it no doubt will continue in the future.

In Chapter 10 we discussed the danger of focusing on scientists associated with the most progress as the "best scientists" and on using them as exemplars of what science is or should be. By defining something based on an extreme, one runs the risk of committing a sharpshooter fallacy, base rate neglect, and observer bias. Yet there is an even deeper problem with using past scientists, great or otherwise, as a defining metric for science. In his book, *Against Method*, Feyerabend basically takes the position that most definitions of science can be rejected because they would result in the labeling of Galileo (and others) as nonscientists.[3] However, Feyerabend was using a modern definition of science that had evolved greatly since Galileo's time and applying it to the actions of a scientist from centuries earlier. This would be equivalent to defining humans as bipedal terrestrial

---

[3] Many others have likewise used similar criteria to assess the validity of attempts to define science.

creatures with the ability to go to the moon and back, to fly from city to city, and to cure many infectious diseases with antibiotics. With this definition, there has never been a human on Earth prior to 1969. However, if one defines humans as terrestrial bipedal creatures with a tendency to make tools, develop technologies, and modify their environment, then the definition can extend back much further.[4] If I were to submit a paper of mine for scientific publication and responded to the reviewer's request for statistical analysis with the comment that Newton, Galileo, and Priestley never did such a thing and thus I don't need to, my paper would never be accepted and I would lose all credibility. This problem can be remedied by making the updating of methods, to compensate for errors as one becomes aware of them, a part of the definition of science.

THE ROLE OF THE PROPERTIES OF THE OBJECT OF STUDY IN DEFINING SCIENCE

One of the major points of this book is that while we can go a long way, if not all the way, in demarcating science from other schools of thought (in my view), science is nevertheless extremely close to normal human thinking. However, in science, normal human thinking is refined through the ongoing development of methods to mitigate errors that humans tend to make. Thus science can go against our natural grain and, as such, feel foreign, counterintuitive, and defy common sense in many cases. This is a hallmark of science that is often missing from other systems of thought, which can clearly be designated as nonscience. However, it also can be correctly pointed out that there are some fields of study, not typically called science, that likewise demand HD coherence and also refine methods over time to decrease error (philosophy, history, etc.). Why are these fields not science?

---

[4] The author takes note that this does not address the issue of whether there is a actual category of "human" in nature as discussed in Chapter 12 and that the current definition does not address all cases.

One argument that we have explored is that science is the study of natural phenomena that fall under the confines of a rule-governed system and to which we have observational access (often including the ability to carry out controlled experimentation of reproducible phenomena) – if not directly, then through some part of the web of belief to which each phenomenon or entity is attached. While other fields may employ the same kind of HD reasoning and general strategy of error mitigation as science, they may vary in their particular topic of focus (e.g., things other than the type of nature that science studies). As such, their ability to test and retest, including the assessment of causal effects through interventional experimentation may be limited or entirely absent. However, this could just indicate a difference in methodological practice as a function of the methods to which the subject of study is amenable, and not a difference in general methodology considered appropriate in an ideal circumstance. Thus, having an object of study with the correct properties so as to be susceptible to scientific methodologies can also be a demarcation criteria. That is, some nonscientists may be acting identically to how a scientist would, but cannot carry out science because of the nature of what they study. In this view, science is not a property solely of the methods of the inquisitor, but is a joint property of the examiner and the examined.

The question of the properties of that which is studied can explain the seemingly odd designations of "hard sciences" vs. "soft sciences" and of sciences that study historical events vs. those that analyze real-time, reproducible phenomena. Different systems are amenable to distinct methods of study and have distinct types of errors. As such, under a common theme of refining methods to mitigate error, the actions of scientists in these areas will appear different even though they both fit under this common theme. Consider the differences in simplicity, consistency, and the ability to carry out controlled experimentation between particle physics, astronomy, human psychology, and human sociology.

Setting the above considerations aside, it has also been argued that these kinds of fine distinctions, called "territorial demarcation," whether an achievable goal or not, is not a problem we really need to address. First, we can distinguish hard science from soft science from certain types of nonscience (e.g., philosophy and history) as a property of their objects of study with a common methodological approach of increasing coherence and mitigating error. Second, there is considerable overlap between the fields, as pointed out by Maarten Boudry: "In philosophy, abstract reasoning and logic take the foreground, whereas in science the emphasis is on empirical data and hypothesis testing. But, scientific theories invariably rest upon certain philosophical underpinnings, and science without abstract reasoning and logical inferences is just stamp-collecting."[5] As Boudry argues, territorial demarcation may also be practically irrelevant. Rather, the real danger is confusing that which has "the look and feel" of authentic science, but is not, as being a legitimate science. Part and parcel with the primary goal of this book, to allow the reader to understand the nature of and how much confidence we should place in "scientific" knowledge claims, is the equally important and related issue of recognizing whether knowledge claims that purport to be scientific really are.

Arguably, any system that makes specific knowledge claims about the natural world under the label of science needs to allow itself to be assessed by scientific standards, and if it will not (or cannot) do so, should abandon the label of science. The basis for differences between history, philosophy, and science is an interesting debate, but historians and philosophers don't typically claim to be scientists, nor do they claim that their scholarly products are scientific findings. This leads to a necessary consideration of

---

[5] Boudry M. 2013. "Loki's Wager and Laudan's Error: On Genuine and Territorial Demarcation." In Pigliucci M, Boudry M (Eds.). *Philosophy of Pseudoscience: Reconsidering the Demarcation Problem.* pp. 79–100. Chicago and London: University of Chicago Press.

pseudoscience as an entity, which is nonscience that makes a specific claim of being science.

Some pseudosciences are incapable of scientific exploration due to how their theories are framed or the focus of their study (or both). They nevertheless insist on being recognized as science (e.g., intelligent design theory – as discussed later). In contrast, other pseudosciences (e.g., astrology – as discussed later) have theory and focus of study that is entirely susceptible to all manner of scientific investigation. However, the practice of the field of astrology refuses to acknowledge or employ methods of science, despite the applicability and utility of such methods. The question of demarcation in these cases is critical. If we cannot distinguish science from pseudoscience, we find ourselves awash in a sea of confusion, meaningless systems of belief, and even "alternative facts" and propaganda.

SCIENCE, NONSCIENCE, AND PSEUDOSCIENCE

Much has been written about pseudoscience, from encyclopedic catalogues of various pseudosciences[6,7,8] to detailed indictments of pseudoscience by scientists and public intellectuals. At the same time, some pseudoscientists (and others) have engaged in a sustained assault on academic thought through the anti-intellectualism movement. As so much has been written on this topic, and since the current work is meant to detail the strengths and weaknesses of science itself, it is outside the scope of this book to explore fine details of pseudoscience in any great breadth. However, to the extent that science can be defined by juxtaposing its properties to

---

[6] Williams WF. 2000. *Encyclopedia of Pseudoscience*. New York: Facts on File.

[7] Shermer M. (Ed.). 2002. *The Skeptic Encyclopedia of Pseudoscience*. Santa Barbara: ABC-CLIO.

[8] Regal B. 2009. *Pseudoscience: A Critical Encyclopedia*. Santa Barbara, CA: Greenwood Press.

characteristics of what is not science, then a brief analysis of pseudoscience is indeed useful.[9]

Moreover, this question is of profound social importance, because the label of "science" allows certain topics to be taught in public schools, often opens up opportunities for funding and support, and lends credibility to various products and services. In addition, as stated elsewhere, Western society seems to recognize the unprecedented progress that science has made in technological development, if not in understanding, and thus those who wish to be accepted are highly motivated to gain the coveted "scientific label."

Nonscience makes use of its own methods, which may differ from those employed in scientific discourse. Of course, there are many ways of understanding the world that are not science and that are happy not to be science. The Romantics felt that the rational enlightenment was no way to live life and a likely road to misery. Many authority-based systems and religions don't want to question their beliefs (at least not in a scientific way as we have defined it), as they find the very act of faith to be meaningful. For them, "science-like attacks" on belief are to be avoided. Individual personal experience may be the currency of discourse, and there is no need to question observation in a fundamental way nor to carry out controlled experiments, isolate variables, or apply advanced statistics and probability theory; it doesn't factor into the belief system. This is not to say that people in these areas are not intelligent, thoughtful, and scholarly. It is just an approach that does not line up with our working definition of science.

Most nonsciences are happy to use their own methods and certainly have no ambitions to appear to be scientific. In some cases, they take great pride in using a different approach to understanding the world. In contrast, pseudoscientific thinking is clearly not science but has the "look and feel" of scientific systems.

---

[9] In recent years, much of the "demarcation debate" has taken place in the context of juxtaposing science with pseudoscience. Pigliucci and Boudry, 2013.

Pseudoscience has its own theories, terminologies, retroduced entities as causes, observed phenomena as effects, systems for both explaining the past and predicting future events and, in some cases, specialized instrumentation to measure causes or effects that are beyond the ability of normal human senses to detect. Pseudoscience sometimes has societies that develop theories, hold meetings with presentations, and encourage discourse, as well as journals that publish findings, some that even have a peer-review process. This sounds an awful lot like the aspects of science we have described in this book, which is exactly the point. Pseudoscience has all of the trappings of science and yet is nonscientific. In some cases, the motivation for adopting scientific characteristics is the desire to be accepted as science, sometimes in a disingenuous manner[10]; in other cases the motivation is a sincere belief in the system. While it is not useful to drag out a litany of things that may be labeled as pseudoscience, a few examples are necessary as a means of illustration. Popular examples will be used, both because they may be familiar to the reader and because their analysis has been well developed by others.

An important example of pseudoscience is intelligent design (ID) theory as it relates to the origin of species on Earth. Although ID does not require a specific deity to be responsible for Earth's creation, it states that an "intelligence" of some sort created the species all at once and as they exist today. ID is often placed in the boxing ring the theory of evolution, which raises a number of issues. As explained earlier, the theory of evolution (at least as restricted to

---

[10] At times, pseudoscience will use linguistic tricks to sound as though it has some scientific merit, but without an outright lie. Typically, descriptions are given that, while not an outright lie, purposefully misrepresent what has actually been shown by implying a certain result when one has not been obtained. For example, when advertising a therapy of some type, one might hear a treatment described as having been "clinically tested" or having been "university studied." Note that while these phrases appear to add academic credibility, in fact, they say nothing about what the tests were or how the tests or studies were run, nor do they comment on the outcome of the test or studies.

the origin of species as catalogued by humans) is a historical science, and while this is still science, it nevertheless has a different character than systems in which one can actively experiment today.[11]

Nevertheless, if evolution is indeed a scientific theory, why is ID not a scientific theory? Both ID and evolution make reference to historical events (although ID's events happened over a very short period of time and evolution's events spanned millions of years). Both ID and evolution hypothesize a process by which living things came to be in their current state (in the case of ID, the designer made them; in the case of evolution, life started simply and slowly developed over time, based on natural selection acting on variations through random mutations). Both ID and evolution acknowledge the same fossil record. In the case of ID, the designer put the fossil record there; in the case of evolution, the record formed over millions of years through the fossilization of dead animals.

ID and evolution are entirely compatible in many ways, as the theory of evolution is silent on how the whole "life thing" got started. While many theorists of evolution would posit a spontaneous chemical reaction leading to self-replicating polymers that gave rise to life in primordial mud puddles on an ancient Earth, there is no reason that an intelligence couldn't have given the spark of life to the first forms and then allowed them to evolve. Indeed, evolution could even be the purposeful instrument by which the intelligence intended life to gain its complexity. However, it is the insistence that all life came to exist simultaneously and with its current complexity, and not as a gradual process of slow modification through random mutation and selection, that puts ID theory at firm and complete loggerheads with the theory of evolution.[12]

---

[11] Importantly, there is a great deal of experimental science going on today that focuses on evolution happening right now and things other than retroducing how diversity of species came to exist on Earth (through either evolution, ID, or some other source), which is the debate upon which ID focuses.

[12] Although ID does not state any specific deity, it is an outgrowth of Creationism, which comes from the Abrahamic religions and is based on the Bible. As such, there is an underlying focus on maintaining an age of the Earth of approximately 6,000

ID has almost all of the necessary components of a scientific program. There is a theory with base hypotheses, one can predict a certain outcome from the theory, and one can empirically test if the outcome is in fact observed (i.e., a great diversity of species). Also, like hard-core sciences, ID has societies and foundations devoted to its study, institutes and intellectual groups, and journals and vehicles of publication that have at least some form of peer-review. Why, then, is ID not a science by our definition? The reason is not because one can never prove the existence of an intelligent designer. Indeed, science cannot "prove" the existence of unobservable scientific entities. Just as, for example, no one can prove the existence of electrons. We can only *observe* effects consistent with electrons (the problems of retro-duction and underdetermination from Chapters 2 and 3, respectively).

The reason ID is not a science is that the outcomes are not deducible from the premises. That is not to say that one can't predict the outcomes, i.e., "My hypothesis is that there is an intelligence that designed multiple species and, low and behold, I observe just that." The problem is that an intelligence can do whatever it chooses to do. If one points to the fossil record and states that all species should appear together in the fossil record if they were generated at one time (which they do not), the ID defender could simply reply that the intelligence created the fossil record in the way that it is. Because the intelligence is a free cognition, it can do different things given the same initial conditions and identical auxiliary hypotheses – it can "work in mysterious ways" and can have capricious whims. The outcome of ID is not deducible, meaning that an outcome must occur from a given hypothesis and auxiliary hypotheses. Without this characteristic, the minimum requirements of science cannot be met. Again, if a hypothesis can never be rejected under any circumstances, even if one grants that all other things are equal (e.g., auxiliary

years based on the Old Testament. This is the main reason ID cannot allow for an intelligence to have created simple life and then had the life evolve into the current diversity of species over millions of years, as this violates the age of Earth that is indicated by Biblical scripture.

hypotheses are fixed), then no science can be done with it. It is the same problem of being a web of belief, but not a HD web of belief, as was explored with the Seekers in Chapter 4. It is a case in which a supernatural being is not rule governed as was discussed in Chapter 5. The Seekers were not a pseudoscience because they never presented themselves as scientists nor claimed any scientific status; ID demands recognition as a science.

The argument against ID being a science by no means depends upon the theory of evolution being correct or incorrect; the objection to ID stands on its own. Certainly, the theory of evolution is not perfect (as no science is perfect). Indeed, consistent with the best scientific thinking, Charles Darwin devoted an entire section of his book on natural selection to the problems with his theory. His doing so is highly characteristic of good scientific work and self-skepticism. The fossil record does not support a continuous progression of change; rather, it can go in fits and spurts, necessitating modification of evolutionary theory to include punctuated equilibrium; thus, like most good science, the theory had to be modified as new and unpredicted data were discovered.

In contrast to evolution, ID can handle any data whatsoever, and without modification of any part of its web of belief , based on the phrase, "Well, that's what the designer decided to do." In the event that new excavations found all species coexisting in the same paleontological stratum, or if geologic theory changed such that fossil records would appear stratified over time even if all the species existed simultaneously, then evolutionary theory would really have a scientific problem, leading to the need for a profound modification of the theory, or its outright rejection. In contrast, there are no data – none whatsoever – that could compel ID to change or lead to its rejection.[13] Thus, the nature of ID theory makes it incapable of being assessed scientifically, and as such, it cannot be science.

---

[13] One could argue that ID predicts that Earth is only 6,000 years old, and that measurements of the Earth's age that show it to be much older reject ID. In some

Much of the academic energy of ID proponents is spent in attacking the theory of evolution. I find this highly commendable. To their immense credit, proponents of ID theory are actually acting as scientists when they point out contradictions between what evolutionary theory predicts and what the fossil record contains. However, although this may be a scientific evaluation of evolution, it has nothing to do with a scientific evaluation of ID theory. Even if natural selection and evolution were rejected as a theory, this would not constitute proof of ID. The idea that only evolution or ID can explain the diversity of species, and that if evolution is false, ID must be true, represents the fallacy of a false dichotomy (also called the fallacy of limited hypotheses). Rejecting A can only prove B to be true if A and B are the only possible theories that can explain something. Clearly, there are more than two theories that can explain the diversity of species. Indeed, due to the nature of retroduction and affirming the consequent, there are endless theories in addition to evolution and ID that can explain diversity of species. Attempting to prove ID by ruling out an infinite number of alternate theories is clearly not possible. Stating that ID is true because evolution is false is like stating that all ravens are black because I saw a lavender apple. It is not zero evidence, but it is pretty close.

Ironically, when scientists object to ID theory because they don't think it predicts natural phenomena, they are not making a scientific objection when they do so, as ID makes no deducible predictions. It doesn't even qualify as bad science, it cannot fulfill the minimum requirements of a theory that is assessable by scientific methods. Of course, these objections don't make ID theory wrong, nor does it invalidate the study of ID theory as an activity;

cases ID proponents argue that the methods of measuring the age of Earth are not accurate, and this, by the way, is highly consistent with good scientific practice. However, ID can concede that the measurements are correct, but simply evoke the idea that the intelligence made the Earth to look as though it was millions of years old, when it was in fact only 6,000 years old, and this retreats back into a nondeductive system due to capricious whims of a cognition.

it just makes ID not science, in this case a pseudoscience. Just putting on a baseball uniform doesn't make you a baseball player, a lesson I repeatedly learned throughout my youth and that is no less true today.

A second example of pseudoscience that deserves careful consideration is astrology. Although there are many variations on astrology, the central tenet of the belief is that heavenly bodies throughout the universe have an effect on the lives and personalities of individual people. This is an ancient idea, that the Sun and the Moon and the stars affect life on Earth. This idea, in of itself, is by no means absurd nor does it violate some central tenets of the current web of belief of hard-core physics; indeed, there are few if any scientists in general, or astrophysicists in particular, who would deny that the tides of the ocean are a direct result of the gravitational influence of the Sun and the Moon on the waters of Earth. There is firm and existing evidence of celestial bodies altering terrestrial events and even life itself, as many aquatic species have life cycles that are affected by the tides. So what, then, is the problem with astrology?

The problems of astrology are distinct from those of ID. The theory of astrology is an extremely complex and intricate system that links essentially all of the known celestial bodies (Sun, moon, planets of our solar system) and stars directly to the day-to-day events in individual people's lives. Stars are grouped into "constellations" based on certain patterns that have been observed in the night sky. The position of these constellations, the planets, and the Sun at the time of a person's birth is supposed to inform much about their personality and fate, guide them on when the person they marry should have been born, what will happen to them over time, and even what may occur on a given day. Moreover, astrology holds that celestial movements affect all of us in general; for example, when Mercury is in retrograde (appearing to move backwards from its normal path), then general astrological theory holds that the world gets a bit messed up compared with its normal workings.

Astrological charts can be compiled for a given person, and predictions can be made regarding their life in general, and even on a week-by-week basis. To its credit, astrology has many of the components we have identified as being required to be a science. There are causal entities (celestial bodies and constellations) and from them, based on theory and a web of belief, one can predict specific events that are then observable by experience. So, if one chose, one could test if the predictions are correct, and then further evaluate and modify the theory, with self-correction and refinement over time. Sounds like science, right?

I would like to give full credit to Massimo Pigliucci here, who wrote one of the most pointed, concise, and well-developed critiques of astrology and why it is not a science, in his excellent book, *Nonsense on Stilts: How to Tell Science from Bunk*.[14] Pigliucci articulates a number of problems with astrology that we should explore briefly here. Over 65 separate studies showed essentially no correlation of astrological charts with either the personality profiles of the people they were designed for or even agreement with each other. In other words, given the same data, every astrologer came up with a different chart, and none of them correlated with observation better than random guessing. This is a concern, because it questions whether astrological theory is really HD in nature; in other words, if an outcome is deducible from a hypothesis, then the same (or similar) outcome should be deduced by different people who accept the same hypothesis – all other things being equal. This is not to imply that scientists all predict the same thing; such is clearly not the case. For example, when Einstein's theory of relativity predicted that light would bend around strong gravitational bodies, many physicists didn't believe the theory. But, had they believed and understood the theory, they would have

---

[14] Pigliucci M. 2010. *Nonsense on Stilts: How to Tell Science from Bunk*. Chicago: University of Chicago Press.

made the same prediction. It wasn't that Einstein was being illogical; he just forwarded a premise that most others didn't accept (at first).[15] Unlike astrology, scientists are internally deductive. If you give them the same premises, the same rules, and the same background assumptions, they will make the same predictions, at least mostly. In contrast, given the same person to analyze, five astrologers will (on average) come up with five entirely different sets of predictions.

So, based on these results, either astrology is nondeductive, or each astrologer holds a different hypothesis and web of belief about how celestial bodies affect terrestrial events. This latter situation could certainly be the case. Perhaps astrology is a HD science, but there are many different variations on the hypothesis of how celestial bodies affect terrestrial events, and each practitioner has a slightly different hypothesis. Many legitimate sciences have multiple competing hypotheses being developed by different scientists. Let's grant that astrology theory is sufficiently complex, with enough competing versions of the theory so that different practitioners will give very different readings to the same individual. If this is what's going on, then some versions of astrology should predict the specifics of a person's life better than others, and by this, one could start to identify which versions of the theory work best (are most coherent with empirical data) and then refine those versions with ongoing study. However, when a rigorous double-blinded trial was run using 30 of the most famous astrologers, the ability of an astrologer to pick the right astrological chart for any given person was the same as expected by chance alone. In other words, it wasn't better than random guessing.[16] If there is any validity to the theoretical underpinnings of astrology, then at least one of the variations on the theory should predict the

---

[15] That light would bend around strong gravitations bodies was not the premise itself, but was a prediction one could deduce from the premises Einstein put forward and based upon the theory.

[16] Carlson S. 1985. "A Double-Blind Test of Astrology." *Nature* **318**: 419–25.

specifics of a person's life better than random guessing; however, such appears not to be the case.[17]

Pigliucci also points out that in addition to its failure to predict, astrology has poor coherence with the current web of belief regarding celestial bodies. At the time astrology was first divined, it was reasonable to assign identity to the constellations because they appeared as certain shapes. What has since been added to the web of belief is that every star is a different distance from Earth, with large variation in their distances. In other words, although the stars in the sky appear to be two-dimensional on a flat screen of black, they are in fact three-dimensional,[18] and, as such, constellations do not exist (at least they don't have the shape and structure by which astrologers assigned them significance). Moreover, using telescopes, we now know that there are many more stars than astrologers have ever accounted for. Shouldn't these also affect life on Earth? But they are typically not considered by astrologers. As also pointed out by Pigliucci, the stars that are considered appear to exert equal influence regardless of distance. Since none of the known forces of the universe have this property, astrology would have to evoke a new and previously unobserved fundamental force to make this part of the theory plausible. The need of such a force is not in of itself impossible, but it does strain at the existing web of belief and thus would require some evidence.

So, we now understand that astrology is poorly consistent with the web of belief regarding celestial bodies, is highly inconsistent in the predictions it makes from astrologer to astrologer, and that its ability to predict observable outcomes is nil. But why isn't astrology

---

[17] Of course, one can never test all practitioners of astrology, and thus one cannot rule out that some version of a theory has predictive power; however, as the esteemed adherents of astrology were tested, this is probably as good as it gets.

[18] In actuality, the constellations are four dimensional, because the light from each star takes a different time to reach us. Stars may cease to exist but it takes so long for their light to reach us, we don't know this yet. Other new stars may have formed, even a long time ago, and we have not seen them yet. This "time delay" is different for each star, and as such, we don't know exactly what stars exist now.

just "bad science" as opposed to pseudoscience? Why can't it just be a scientific theory that doesn't work?

The reason astrology is pseudoscience is that those who practice it entirely eschew the methods that science has developed to mitigate sources of errors in observation. Many practitioners of astrology, those who do readings and those who receive them, have experienced and observed an amazing predictive power in astrological readings; in fact, it can be downright eerie. This is precisely why well-trained scientists applied modern methods, using trials that were randomized and blinded, a method that is specifically designed to overcome the human tendency to be fooled by their own observation bias. The referenced studies led to the conclusion that when sources of bias are controlled for, then astrology has zero power to predict nature. As such, it is reasonable to conclude that the appearance of astrology as being able to predict events is simply due to well described biases typical of human minds. Any predictive power or validity of astrology is just an illusion.

It's hard to argue that adherents of astrology are entirely ignorant of the scientific findings unless they choose to be, because dozens and dozens of these studies have been published. If they are unaware of this work, they are clearly not in the habit of seeking out information (e.g., investigating the web of belief outside of their narrow myopia). Rather, the adherents of astrology prefer methods that lead to the observation that astrology really works, even if the methods are known to be highly susceptible to error. They prefer anecdotal evidence and processes of personal bias over controlled trials that mitigate known sources of human error. They are unconcerned about the discordance of the tenets of astrology with the web of belief of astrophysicists. They do not mind that every practitioner gives a different reading despite what appears to be a similar theory and that the predictions therefore appear nondeductive in nature (which would explain why they have no predictive value). Professional scientists who have investigated astrology with refined techniques have concluded it's a bad theory that can be easily rejected and have moved on; believers in astrology maintain that it's a wonderful theory, despite and not because of scientific analysis. They prefer other types of analysis and this is the reason that astrology is not a science, at least as it is practiced by astrologers.

When we say astrology is a pseudoscience, we are referring to how the field of astrology and its practitioners interface with the world and the methods they use. They do not favor methods that mitigate error, rather, they favor methods specifically known to result in greater error. Why would one favor methods that are known to be error prone vs. methods that decrease error? This is anathema to science. They likely favor their own error-prone methods because when scientific methods are used, then "the magic" disappears, and who would choose to live in a less magical world? Of course, there is nothing wrong with someone seeking the life experience they prefer. If experiencing the power of the uncanny accuracy of astrology is meaningful to people, if it provides them with the kind of world in which they would rather live – with mysterious forces they can harness and use – this certainly sounds intriguing, if not downright fun. But, such a person cannot embrace a scientific analysis, as doing so entirely destroys the experience they are seeking.

To be clear, the experience of astrology is very real. Practitioners have the experience of highly accurate predictions derived from the forces of the universe. When someone tells you "I know astrology works because I have seen it work, over and over again" they are not lying. They have "seen" it work. However, this real experience is simply a misperception that is erroneously linked to a nonexistent phenomenon by well-described human biases of observation and confirmation. Thus, professional astrologers show a preference for seeking illusion and error rather than decreasing it. It is the opposite of science in this regard. Indeed, if astrologers accepted the methods and approaches that science has developed to mitigate error, then there would be no astrologers or astrology. Thus, astrology as it is practiced does not qualify as a science, not even as bad science, as it seeks and embraces illusion and self-deception. Because it purposely cloaks itself in the trappings of science, this makes it a pseudoscience, one to which the lay public pays enormous sums of money and uses to make important life decisions. They are paying someone to guess randomly for them and then possibly altering their life based on the guess. The only correct prediction a professional astrologer can make is that you will have less money when you leave than when you arrived.

CAN SOMETHING BE A SCIENCE IF IT MAKES NO TESTABLE PREDICTIONS?

I have repeatedly made the point that if a theory does not make at least one testable prediction, then there is no basis for ongoing assessment of coherence between theory, deduction, and observation, failing a minimal standard for allowing a scientific analysis. Some theories are retroduced from an existing amount of data (observations), such that the theory predicts that which has already been observed; however, if one cannot deduce any additional predictions that would lead to new observable outcomes, then one cannot assess the theory further. It is a fallacy of circular reasoning to use the same data from which a theory was first retroduced as confirmatory evidence to support that same theory. To make any progress in assessing a theory, it must lead to a new prediction that can be tested.

When Einstein first introduced the theory of relativity, one of the greatest intellectual achievements of all time, it was not clear that it was a scientific theory because there was no prediction that it made that humans were capable of testing at that time. If the theory was incapable of making predictions, then it wouldn't have any potential to be treated scientifically at all. That was not the case. Many predictions could be deduced from relativity theory, but the circumstances and/or technology didn't yet exist to test them. Eventually, scientists invented instruments or found situations in which they could test the seminal predictions of relativity theory, which no one prior to Einstein would have predicted (e.g., light bending around strong gravitational fields).

If I were to posit a new force in the universe called AZ-waves, but the nature of AZ-waves is such that they can never be measured, nor can the effects of their existence ever be observed, then no science can be done in this framework. Although Einstein proposed special relativity in 1905, many of its predictions could not be tested until years later,[19] as the technology didn't exist or one had to wait for special circumstance (e.g., Sir Arthur Eddington's expedition to the west coast of Africa to observe the solar eclipse on May 29, 1919) to test if light from distant stars bent in response to the gravity of the sun. Are we then comfortable stating that Einstein was a wonderful abstract mathematician and theoretical thinker (one might even call him a philosopher) when he formulated relativity theory, but not a scientist? Are we willing to state that, when first conceived, relativity was not a scientific theory, and only made the transition from philosophy to a science when the technology was invented to test its predictions?

What about the cutting edge theories of physics today? An exciting and innovative theory regarding the intersection of quantum physics and gravitational theory is string theory. As very aptly stated by Massimo Pigliucci, "It is so elegant an idea that it deserves to be true."[20] But is it? Therein is the problem: at the moment, at least, string theory does not seem to make any empirically testable predictions that both differ from those of other competing theories and that can conceivably be evaluated in actual experiments with current technologies.[21] If scientific progress on testing a hypothesis requires

---

[19] As recently as 2010, new technologies have allowed the testing of predictions that followed from Einstein's theories of relativity published almost a century earlier.
[20] Pigliucci, 2010.   [21] Pigliucci, 2010.

that a hypothesis make empirically testable predictions (other than what we already have observed and from which it was retroduced), then string theory is, strictly speaking, not susceptible to ongoing scientific evaluation at this time. Are we willing to accept the notion that there is a whole field of physicists working in some of our greatest academic institutions, being funded by the top scientific agencies in the world, with a robust program of dynamic mathematics and innovative ideas, laboring in a field that is not really a science?

This particular issue can be challenging, and the answer seems strange; however, it is worth arguing that in such areas the intent of those developing the theory (and those who would subsequently consider it) has meaning. The fact that the theory put forth by Einstein was untestable at the time was immediately obvious to the physicists to whom he presented it. Indeed, it is they who went forth and devised very clever ways to test the various predictions that relativity theory put forth. It was with the intent of finding a way to test it that relativity was received, and through this much innovative progress was made. It doesn't matter if those who devised the tests favored the theory, or hated it and wanted to prove it wrong. It was the act of finding new ways to test the theory, carrying forth the tests, and processing the results that mattered.

Measurable effects of the hypothesized Higgs boson were predicted by mathematical theory decades before we built a particle collider large enough to test the prediction, which was observed.[22] Because Newtonian physics does not predict the motions of the heavenly bodies in the universe as we observe them, physicists posited the existence of dark matter, an entity that by its very properties cannot be observed by our current technologies. I have little doubt that those interested in string theory and dark matter are trying very hard to make progress either in the theory (to give more predictions that

---

[22] As discussed in previous chapters, due to the problem of underdetermination this is not "proof" that a Higgs boson exists.

might be testable) or to devise instruments to test existing predictions that we can't currently test. Again, although it may seem odd, it is a strong argument that one must consider the intent of a field with respect to an (as yet) untestable theory, regardless of what the field has yet accomplished, to determine if the theory is being evaluated scientifically or not. If efforts are being made to move closer to being able to test a theory, either by developing the theory to allow more predictions or by advancing technologies to allow testing of current predictions, then the theory is being treated in a scientific fashion. Therefore, purely theoretical fields can be science so long as they are developing the theory in a way that could lead to testing its predictions. In contrast, if no such efforts are underway (theoretical or technological), it is not clear that this is science. Because of the holistic properties of the natural world, technological progress in an unrelated area may inadvertently invent a technology that is capable of assessing a previously untestable prediction, or a seemingly unrelated theory may connect through the web of belief in an unplanned way. Thus, a field may be thrust from nonscience into science due to no fault or credit of its own.

A REALISTIC VIEW OF SCIENCE

In describing science in this work, the attempt has been made to "deflate" some of the exaggerated claims and grandiosities that have been attributed to the scientific program, both from without and within science. The path to disappointment and feelings of betrayal starts with unrealistic expectations. It is hoped that this book has afforded a more realistic view to those who are unfamiliar with the inner workings of science, such that they understand the basis by which scientific claims are made, why they sometimes turn out to be incorrect, and how much confidence to put in them. Scientific claims are not perfect and will sometimes turn out to be wrong. We are always learning as we go. However, the fallible nature of science does not mean that a scientific claim doesn't have a different character than a nonscientific claim. It does not mean that we should accept all knowledge claims as equivalent. It does not mean that we should accept "alternative facts"[23] as being on equal footing with facts based on and consistent with rigorously obtained and analyzed evidence. It does not mean that we should attribute the same properties to pseudoscience as science. We should accept scientific knowledge claims for what they are, as claims defined by certain properties and with particular limitations, but nevertheless using a system that has consistently made more technological progress and better predictions of the natural world than any other system known to humans.

Although such may have been the case in the past, most modern scientists don't typically consider themselves to be seeking a deeper underlying truth. This is for several reasons, which have become clear as science has matured. The error of affirming the consequent, which is present in all retroduction (performed both as part of science and otherwise) leads to an underdetermination that prohibits "knowing" whether a retroduced thing really exists or a retroduced cause is actually true; that is, without solving the entire universe, which it seems we are not exactly at risk of accomplishing anytime soon. As humans are quite capable of inventing abstract and metaphysical notions that may only exist in our minds, we have an endless source of underdetermination that exceeds the content of the natural world outside our cognition. This is the problem being addressed by Occam's Razor, that one should not invent another unobserved cause without it being needed, as one could go on this way forever (indeed, some systems of belief, typically nonscientific, have happily done this).

Even if we were able to limit ourselves to the natural world and its "actual" contents, our ability to observe nature is both finite and flawed. We will never know all of the "facts" of nature, and we will always assign "factual" status to some things that really aren't the

---

[23] The term *alternative facts* was coined by Kellyanne Conway in 2017 on the January 22 episode of "Meet the Press" in an attempt to justify a statement that was clearly incorrect when compared to existing evidence. It is a euphemism for an incorrect conclusion that one refuses to admit is wrong.

case, and so our web of belief will always be flawed in some way. Over time we will continue to modify the web in light of new observations and theory, always seeking greater degrees of coherence, and in doing so hopefully arrive at more and more useful hypotheses and understanding. Indeed, when scientific revolutions occur, and paradigms shift, we may even fundamentally change the web we are studying, swapping certain utilities and limitations for others. For each of these reasons, while ongoing progress seems likely, it is guaranteed that any completion of the web and its reconciliation with reality will remain elusive.

Error can take place at each of the critical steps. The natural world can be misobserved and misinterpreted, retroduced hypotheses may not lead to deducible predictions of already known phenomena, and new predictions may be improperly deduced from the hypothesis. New approaches or instruments may need to be invented to test predictions, and errors can be made in such tests, even if the technology already exists. Observations may be flawed due to the failure of instruments or human senses. Chance effects may cast doubt on our ability to observe associations. Studies may not be controlled properly, and all manner of normal human bias may be inadequately compensated for. Data that appear to support a hypothesis may be favored and data that go against a hypothesis underemphasized, due to lack of controlling for human confirmation bias and special pleading. Even if rejecting data are acknowledged, any hypothesis can be rescued by changing an auxiliary hypothesis. Alternatively, data that seem to go against a hypothesis may be exaggerated by those who have a preexisting bias against the hypothesis. Even if our understanding of past events were perfect, problems with induction make it certain that we will make errors in trying to predict the unobserved based on the observed.

In short, science is a terribly flawed enterprise, with numerous sources of error at multiple levels. However, this somewhat deflated view of the grandness that is often attributed to science, this vision of real scientific process (warts and all), should not distract us from the fact that scientific practice is fundamentally different from other approaches to the world, that science can be defined and demarcated from nonscience and the pseudoscience that wraps itself in the cloak of scientific appearance.

It must be acknowledged that there is no promise that science has or will move us toward an abstract truth or that it will bring us closer to knowledge of a truth that we may never fully achieve. Yet with regard to increasing our ability to predict and control the natural world, no human enterprise has achieved the progress that scientific methods have. In fact, none has even come close. Predicting that science will continue to achieve progress suffers from the uncertainty of induction but my bet is that it will do so, as long as its focus remains on the never-ending quest to better understand our sources of error in observation and thought, and to refine our methods to mitigate such errors, as we tool-making primates, who are regrettably flawed, charmingly persistent, indefatigably curious, and often confused, continue to explore the natural world.

About the Author

James Zimring, MD, PhD, carried out his academic training at Emory University, where he was awarded both a PhD in immunology and a medical doctorate, followed by a residency in clinical pathology with a focus on transfusion medicine and a postdoctoral fellowship in cellular immunology research. He is a diplomate of the American Board of Pathology with certification in clinical pathology.

In 2002, Dr. Zimring was appointed as assistant professor at Emory University, followed by promotion to associate professor with tenure in 2007. In 2012, Dr. Zimring moved his lab to the Bloodworks Northwest Research Institute (a not-for-profit research institute focusing on blood biology), where he was appointed as a full member, and subsequently as chief scientific officer. In 2013, Dr. Zimring was appointed as professor in the Department of Laboratory Medicine and later in the Department of Medicine (Division of Hematology) at the University of Washington School of Medicine. In 2019, Dr. Zimring was appointed as professor of pathology at the University of Virginia in Charlottesville, where he continues to develop and pursue his research, including studies in transfusion biology in general, with a focus on blood storage and immunology.

The author of numerous articles and the recipient of many awards for his research and teaching, Dr. Zimring is recognized as an international expert in the field of transfusion biology and routinely delivers academic lectures both nationally and internationally.

Index